# TWO TOP PRIORITY PROGRAMS TO REDUCE UNEMPLOYMENT

Lift Wage Rates To Expand Consumption
And To Catch Up With Productivity Gains

Launch A Much Larger Housing Program
To Help Counteract The Elimination Of Jobs
Caused By Technology And Automation

CONFERENCE ON ECONOMIC PROGRESS

1001 Connecticut Avenue, N.W.    •    Washington 6, D. C.

DECEMBER, 1963

# Contents

Page

I. Summary And Theme: Both Private And Public Action
Are Needed To Restore Full Employment................................... 1
We are moving backward on the unemployment problem.............. 1
The trouble within the private enterprise area .... .... ... ........ 1
The central importance of the wage deficiencies ..... ..... ........... 2
The role of the shorter work week........ ... ... ......... ...... ........ 2
The urgent need for large increases in public spending ..... ...... 3
Tax reduction alone is a frail weapon .... .......... ......... 3
Proposed increases in the Federal Budget ... .... ......... ........... 4
A great new housing effort: prime response to technology
and automation .... ..... ............. .... ... 4
The low Federal costs in terms of the benefits ... .... .......... 5
The two basic proposals are consistent and mutually reinforcing ........ 5

II. The Chronic Rise Of Unemployment............................ 7
The true level of unemployment ....... ........ ......... 7
Unemployment is widespread, and hurts the whole Nation .... ......... 7

III. Rising Unemployment Reflects Low Economic Growth.............. 13
The current "recovery" thus far is unsatisfactory ........... ..... .......... 13

IV. The Main Trouble: Deficiency In Demand
Or Purchasing Power.................................................... 15
The dominating deficiency is private consumer spending ...... ......... 15
The deficiency in public outlays for goods and services .......... ..... ...... 15
The problem of business investment .... ........ ......... .... ...... ......... 16

V. The Lag In Wages And The Wage Deficiency............................ 22
The huge size of the wage deficiency ...... ...... ............. 22
Wages have lagged far behind productivity gains ... ...... ........... 22
Appropriate standards for wage rate increases............................ 23
The distinction between actual productivity trends and true trends
in our productivity-potential .... .... ......... ..... ..... ............. 24
Tendency toward acceleration of productivity gains ..... ................ 24
Comparative trends in profits, investment, and wages .... ....... 25
Relationship between wage rate increases and fringe benefits ......... 25

VI. Goals And Programs For Wage Improvement............................ 32
The size of the wage-lift task ...... ..... .................. 32
Needed improvements in minimum wage legislation ........................ 32
The wage problem and the shorter work week............................ 33
The shorter work week would not be inflationary............................ 33
Massive poverty in America is largely a wage problem ...... ........... 34
Other measures needed to improve incomes .............................. 34

# Contents (continued)

Page

VII. The Changing Structure Of Employment Opportunity:
The Problem Of The Structure Of Demand.............................. 37
Significance of the structure of demand ................................. 37
Impact of technology and automation ...................................... 37
Employment trends, production workers in manufacturing ........... 38
Employment trends, all nonagricultural wage and salary workers,
and in the entire civilian economy ...................................... 38
Unemployment rate trends, wage and salary workers .................. 39
The labor force adapts itself well to changes in the structure
of demand ............................................................................ 39

VIII. Employment Goals, Looking Toward 1966 And 1970.............. 45
Difference between forecasts and goals ................................... 45
Projected increases in civilian employment, by sectors ............... 45

IX. The Imperative Need For More Federal Public Spending........... 49
General summation of the need........................................... 49
The specifics of the need.................................................... 49
Tax reduction is no substitute for increased Federal spending ...... 50
Specific defects in the pending tax bill ................................... 51
Proposed revisions in the tax bill........................................ 51

X. Economic Need For Greatly Expanded Housing Efforts.............. 56
The importance of housing in the U. S. economy to date ............ 56
The role of housing in economic restoration ........................... 57

XI. Social Need For Greatly Expanded Housing Efforts.................. 59
Extent of bad housing in the U.S., and the outlook.................. 59
Goals for improved housing conditions................................. 59
The human and public costs of bad housing ........................... 60

XII. Bad Housing Results From Low Incomes................................. 63
Unsound housing related to income levels ............................... 63
Housing conditions related to housing costs........................... 63

XIII. Toward A Decent Home For Every American Family............. 67
Why the need is not now being met....................................... 67
The proposed new effort ...................................................... 68
The moderate Federal costs in terms of the benefits.................. 68
Relationship between housing and urban renewal.................... 69
The farm housing problem................................................... 69
Unusually high "multiplier" effect of Federal aid to housing and
community development ..................................................... 70

# Charts

Page

Chronic rise of unemployment and of idle plant, 1953-1963 ............... 9
Amount by which true level of unemployment has exceeded level
  consistent with full employment .... ...................................... 10
Total of those unemployed shown by category, 1963 ................... 11
Large national economic deficits during period 1953-3rd quarter 1963 ...... 12

Growth rates, U. S. economy, 1922-1963 .......................... 14

Deficient rate of growth in private consumer spending, 1953-1963 ........... 17
Low growth in private consumption reflects even lower growth
  in incomes in most recent years ..... ....................................... 18
Federal Budget has shrunk relative to size of
  economy and needs, 1954-1964 .......................... 19
Investment in plant and equipment was deficient, 1953-1963 as a whole  20
Key profits after taxes are high despite large unused capacities ............. 21

Deficient rate of growth in wages and salaries, 1953-1963 ..... .... ......... 26
Comparative trends in productivity and real hourly earnings, 1957-1962 .. 27
Trends in productivity for the entire private economy, 1910-1962 .......... 28
Before the 1957-1958 recession, profits and investment outran wages—
  basic to consumption .......................................... 29
Before the 1960-1961 recession, despite reduced profits,
  investment outran wages—basic to consumption ..................... 30
During current economic upturn, profits, and investment
  in some cases, outrun wages—basic to consumption ..................... 31

Goals for 1966 and 1970, projected from actual levels in 1963 ............... 35
Americans living in poverty and their share of income, 1961 ............... 36

Ratio of volume of employment to physical volume of production .......... 40
Employment trends: production workers, 1947-1963 ....... ... ......... 41
Employment trends: nonagricultural wage and salary workers, 1947-1963  42
Total civilian employment trends, by occupation, 1947-1962 .... ......... 43
Unemployment rate trends, wage and salary workers, 1947-1963 ............ 44

Goals for total civilian employment, by occupation,
  1962-1966 and 1962-1970 ....................................... 47
Total civilian employment distribution, 1947, 1953, and 1962,
  and goals for 1966 and 1970 ....................................... 48

# Charts (continued)

Goals for a Federal Budget geared to economic growth and public needs  52

Toward a Federal Budget consistent with maximum employment
and the priorities of national public needs  53

Pending tax bill: estimated division between cuts for investment
purposes and cuts for consumption purposes  54

Pending tax bill, personal tax cuts  55

Role of housing in the U.S. economy, 1947-1962  58

Total number of housing units and number seriously deficient,
in 1960 and projected to 1966 and 1970  61

Substandard housing breeds economic and social ills  62

Housing conditions related to incomes in metropolitan areas, 1960  65

Housing conditions related to costs in metropolitan areas, 1960  66

Incomes of purchasers of new FHA single family homes,
1st quarter 1963  71

Toward decent homes for all: goals for new nonfarm housing  72

# I. Summary And Theme:
# Both Private And Public Action
# Are Needed To Restore Full Employment

## We are moving backward on the unemployment problem

Unemployment during the first eleven months of 1963, measured as a percent of those able and willing to work, was almost twice as high as in 1953. Despite an economic "recovery" now said to be one of the "longest" on record, unemployment during the first eleven months of 1963 and in November 1963 was higher than during the corresponding periods of 1962. The true level of unemployment is now about 7 million. About 3¼ million, equating with full-time unemployment as officially reported of about 2 million, would be consistent with the prime national objective of maximum employment under the Employment Act of 1946. This true level of unemployment includes full-time unemployment, the full-time equivalent of part-time unemployment, and unemployment of those not counted as such because the shortage of jobs has discouraged their looking for work.

Unemployment is not an insoluble problem. It can be cured, and cured only, by bringing total demand for ultimate goods and services into line with our increasing ability to produce them under the new technology and automation. Under the American system of responsible free enterprise and responsible free Government, this is a task for both. But we are neglecting both aspects of this task, and failing to recognize the inescapable relationship between the two.

Private enterprise, instead of initiating the steps which it can and should take, is waiting to see what the Government will do.

The Government is permitting an "economy drive" to snowball to such an extent that our native concern about the unemployed, the poor, and the deprived may become frozen in an icy indifference.

## The trouble within the private enterprise area

The main default on the private enterprise side has been failure to maintain a satisfactory balance between the flow of funds into investment in the plant and equipment which add to our productive capabilities, and the flow of funds which add to the ability of consumers to buy enough to keep the plants fully running and the manpower fully employed. This explains how each short period of sharp economic upturn has been con-

1

verted into stagnation, and how each stagnation prior to the current one has been converted into recession. During the current stagnation since fourth quarter 1961, our average annual growth rate has been less than 4 percent, or less than half that required for at least two years to lift us from where we are now to maximum employment and production. And during this current stagnation, profits have soared to higher levels than investors can use; investment in plant and equipment has once again expanded faster than demand for ultimate goods and services; and consumer purchasing power has continued to lag seriously.

## The central importance of the wage deficiencies

For 1963 estimated as a whole, a consumer spending deficiency of about 59 billion dollars has been more than three-quarters of the total national production "gap" of 76.5 billion. Meanwhile, a 56.4 billion dollar deficiency in wages and salaries has been more than three-quarters of the total personal income deficiency of 72.9 billion which explains the inadequate consumer spending.

The chronically growing wage deficiency results largely because the hourly wage rates of those employed have lagged far behind productivity gains or increased output per man-hour worked. Until the huge shortage in purchasing power resulting from this is remedied, adequate reduction of unemployment remains impossible. During the most recent five-year period 1957-1962, while productivity in the whole private nonfarm economy rose at an average annual rate of 2.7 percent, real hourly earnings rose at an average annual rate of only 2.1 percent. In manufacturing, while productivity rose at an average annual rate of 3.4 percent, real hourly earnings rose at an average annual rate of only 1.8 percent. These vast disparities are augmenting even now.

The largest single step toward economic restoration needs to take the form of increases in hourly wage rates, to catch up with productivity gains and to reflect the accelerating rate of productivity gains under the new technology and automation. Starting with 1963 as a base, wages and salaries need to rise (through rate increases and reemployment) by about 72 billion dollars by 1966, and by 160 billion by 1970.

## The role of the shorter work week

Hourly wage rates can be lifted either by maintaining the current length of the standard work week and increasing weekly take-home pay, or by shortening the length of the standard work week to 35 hours or thereabout while maintaining weekly take-home pay. Neither of these alternatives would be "inflationary" in terms of the increased demand for goods and services which they would generate, in view of the immense shortage of such demand now and in the foreseeable future. Neither of these alternatives should be set aside on the ground that it would "force up prices." Profit margins are now too high, and a better balance between funds available for investment and funds available for consumption would expedite economic growth, reduce idle manpower and plant, and enhance aggregate profits and investment in the long run. And in view of the fact that no dent in unemployment with the current standard work week is foreseeable under private and economic policies now under active consideration, a shorter work week appears manifestly desirable.

## The urgent need for large increases in public spending

But adjustments within the structure of private enterprise cannot do the whole expansionary job. First, the size of the job is too big and the need for speed too urgent. Second, the rates of technological advance and automation in various industrial sectors and in agriculture are moving so much faster than any feasible expansion of consumer needs and wants in these areas that it will be impossible to restore maximum employment by increases in purchasing power directed solely at goods and services produced in these areas. A large proportion of the structure of total demand must be redirected to those types of goods and services which require new admixtures of private and public spending, with heavy accent upon increased Federal spending. Third, only this increased public spending can meet some of the top priorities of our unmet national needs. These include, among others, rehousing of slum dwellers and renewal of our urban areas; enlargement of educational and health and recreational facilities and services; improvements in mass transportation; and expansion of resource development and conventional types of public works.

## Tax reduction alone is a frail weapon

The amounts in the now-pending tax reduction bill and the timing of their application can carry at best only a small part of the restorative burden. The types of demand resulting from tax reduction will be too largely directed to areas where the rate of technological advance and automation, for reasons stated just above, will not permit very large or fast additions to employment. Nor is tax reduction very helpful in meeting the top priorities of our domestic public needs. In addition, changes

3

in the tax bill are needed to redirect much more of its benefits to middle- and low-income consumers in order to achieve larger stimulative effects. The mounting pressures to hold the line on Federal public spending in exchange for tax reduction is a very bad bargain on all scores.

## Proposed increases in the Federal Budget

The Federal Budget needs to play its part in helping (a) to attain an adequate volume of total economic expansion, (b) to bring about the needed changes in the structure of demand responsive to the new technology and automation, and (c) to reverse the progressive starvation of essential domestic programs. This study proposes that the Federal Budget for fiscal 1965 (as presented in January 1964) be 8 to 8.5 billion dollars higher than the original Budget for fiscal 1964, or about 107 billion, with 3.5 to 4 billion of the increase allocated to domestic programs.

## A great new housing effort: prime response to technology and automation

This study concentrates upon greatly accelerated housing and urban renewal, as the largest single opportunity to direct increased Federal spending toward solution of the basic problems defined just above. The proposal is designed to activate maximum increases in private employment and private investment for every dollar of increased public spending, and through resulting increases in tax revenues to combine these needed increases in Federal spending with an optimum long-range program to reduce greatly and ultimately to eliminate the Federal deficit.

For 1963, new nonfarm housing starts are estimated at 1.6 million. But practically all of this serves middle- and high-income groups, and so it has been in years gone by. This recurrently saturates the market. It leads to the extraordinary instability of housing starts which contributes to general economic instability. It furnishes inadequate expansion of employment opportunity. It makes only picayune progress toward helping the one-fifth of the Nation who are still ill-housed.

Therefore, this study proposes a new long-range housing effort, very large in size and scope, starting at once and running through 1970. The program is pointed toward two million new nonfarm housing starts in 1966, and 2.2 million in 1970, divided as follows (1) about 1.2 million units in 1966 and in 1970 of conventionally financed private housing for middle- and high-income families; (2) about 0.4 million units in 1966 and about 0.5 million in 1970 for lower-middle income families. This would be privately financed housing, but with Federal assistance directed toward

4

lower interest rates, longer amortization terms, and aid to land assembly and other aspects of urban renewal; and (3) about 0.4 million units in 1966 and about 0.5 million in 1970 for low-income families who now live in slums, with Federal subsidy assistance as well as State and local assistance.

## The low Federal costs in terms of the benefits

The proposed program would require that the item in the Federal Budget for housing and community development (urban renewal) be lifted from about one quarter of a billion dollars in fiscal 1964 to 2.2 billion in calendar 1966 and 3.3 billion in calendar 1970. Even with these lifts, these outlays would comprise less than 3 percent of the Federal Budget in 1970 and only 0.37 percent of what our total national production should be by that year.

Federal outlays for these purposes—3 billion higher in calendar 1970 than in fiscal 1964—would contribute powerfully to a projected increase of almost 23 billion in private residential nonfarm construction in 1970, compared with calendar 1963. This in turn would contribute powerfully to a projected increase of about 40 billion in private investment in total new construction. Coordinately with this, employment in contract construction (of which employment in residential construction should be an increasingly large part), measured from a 1962 base, should and can be lifted 37.8 percent by 1966 and 48.6 percent by 1970, contrasted with a lift in total civilian employment (from a 1962 base) of 13.4 percent by 1966 and 22 percent by 1970.

## The two basic proposals are consistent and mutually reinforcing

The proposal (1) to expand private consumer spending by wage rate increases, and the proposal (2) to increase Federal outlays on the ground (among others) that this is needed to redirect demand toward areas where increases in private demand are not likely to add nearly enough to employment expansion, are entirely consistent.

Even in those areas where increases in private demand will not add very greatly to employment expansion because of the rate of the advancing technology and automation, these increases in private demand will nonetheless help to reverse the downward trend in employment occurring in many of these areas, and indeed promote some sizable increases. Further, the wage increases will purchase goods and services which enter into improved living standards quite as much as public services. In advocating

5

enlargement of the "public sector", we should not go overboard and lose perspective as to the essential nature of our economy, or as to the prevalence of the private low living standards in our midst which are quite as shocking in human terms and quite as inimical to economic growth and full employment as the poverty in the public sector. Large advances are needed on both the private and the public front.

Moreover, these two advances reinforce one another. Better education and health services, when publicly financed, help people to become fitted for higher grades of private employment. Higher grades of private employment, and higher wages for productivity in the same employment, help people to pay the cost of improved education and health services, either in the form of private outlays or in the form of their contributions to tax revenues. In short, the reduction of poverty and deprivation and the lifting of living standards in the United States are one and the same problem as the problem of speeding up economic growth and achieving and maintaining maximum employment and production. A sufficiently comprehensive and effective listing of the programs required for either of these two purposes would turn out to be practically the same as a listing of the programs required for the other.

This study has been directed by Leon H. Keyserling, with the primary assistance of Mary Dublin Keyserling and Philip M. Ritz. Martin H. Seiden and Antoinette Chautemps were also helpful.

# II. The Chronic Rise Of Unemployment

### The true level of unemployment

The actual volume of unemployment is seriously understated by the usual measurement which deals only with full-time unemployment. Even so, full-time unemployment rose from 1.9 million or 2.9 percent of the civilian labor force in 1953 to 4.2 million or 5.7 percent in the first three quarters of 1963.

But there are two other types of unemployment, as shown by the chart on page 9. First, if ten people work only half the time, or only 19 hours a week in a normal work week of 38 hours, this part-time unemployment is equivalent to five people unemployed full-time. Second, many of the people who are discouraged from looking for work because jobs are scarce or unavailable are not counted as being in the civilian labor force, and thus are not counted as being unemployed. This is really concealed unemployment. Adding these two additional types of unemployment to full-time unemployment, the true level of unemployment rose from 3.2 million or 4.9 percent of the civilian labor force in 1953 to 7 million or 9.5 percent in the first three quarters of 1963. This contrast is even more striking because a recession started around mid-1953, while 1963 represents the third year of an uninterrupted "economic recovery."

The chart on page 10 shows the amount by which the true level of unemployment has exceeded the amount consistent with full employment. In 1953, this "excess" unemployment was only 0.3 million; during the first three-quarters of 1963, it was 3.7 million, or more than twelve times as high. Comparing the same two periods, the average duration of unemployment per unemployed worker rose from 8.1 weeks to 14.2 weeks; those unemployed 15-26 weeks rose from 7 percent to 13.1 percent of all those unemployed; and those unemployed 27 weeks or longer rose from 4.2 percent to 13.6 percent of all those unemployed.

### Unemployment is widespread, and hurts the whole Nation

During the first nine months of 1963, as shown by the chart on page 11, of all those unemployed 25 percent were in manufacturing. But 16.8 percent were in wholesale and retail trade, and 15.3 percent in the service industries. This punctures the notion, common a few years ago, that those forced out of manufacturing would find jobs in the so-called white collar occupations. Persons with no previous work experience accounted for 15.3 percent of total unemployment (measured against the size of this group, unemployment among those 16-19 years of age was three times

7

as high as the nationwide average). And 11.4 percent of total unemployment was in construction. Thus, unemployment is very widespread, not limited to a few sectors of the economy.

Others besides the unemployed are hurt by high unemployment. As shown by the chart on page 9, the deficiency in our total national production rose from 1.5 billion dollars or 0.3 percent of maximum production in 1953 to an annual rate of about 76 billion or 11.6 percent in third quarter 1963. And for the period beginning with 1953 through third quarter 1963, as shown by the chart on page 12, the same conditions which caused a loss of 28 million man-hours of employment opportunity caused a loss of 475 billion dollars in total national production, 115 billion in private business investment opportunity, 71 billion in farm operators' net income, and $7600 in the income of the average family.

The four following charts illustrate this chapter.

# CHRONIC RISE OF UNEMPLOYMENT
# AND OF IDLE PLANT, 1953–1963 [1]

## TRUE LEVEL OF UNEMPLOYMENT
### (Millions of Workers)

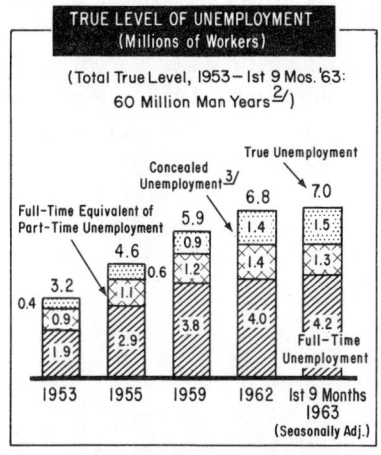

(Total True Level, 1953 – 1st 9 Mos. '63:
60 Million Man Years [2])

1953    1955    1959    1962    1st 9 Months 1963 (Seasonally Adj.)

## UNEMPLOYMENT AS PERCENT OF
## CIVILIAN LABOR FORCE [4]

1953    1955    1959    1962    1st 9 Months 1963 (Seasonally Adj.)

## DEFICIENCIES IN G.N.P.
### (Billions of 1962 Dollars)

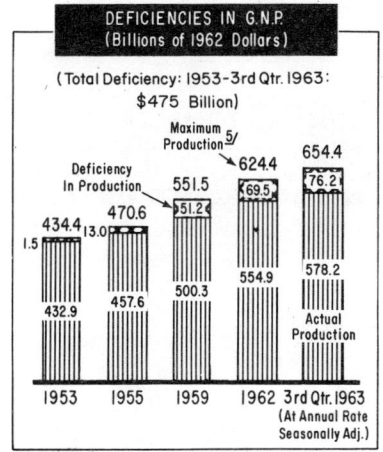

(Total Deficiency: 1953 – 3rd Qtr. 1963: $475 Billion)

1953    1955    1959    1962    3rd Qtr. 1963 (At Annual Rate Seasonally Adj.)

## DEFICIENCIES AS PERCENT OF
## MAXIMUM PRODUCTION

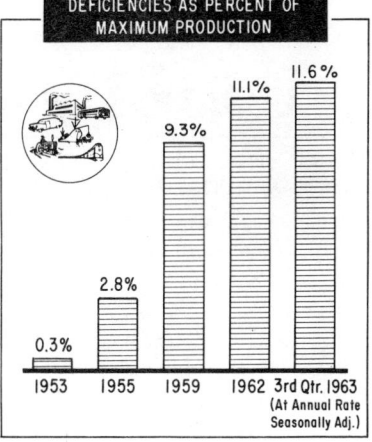

1953    1955    1959    1962    3rd Qtr. 1963 (At Annual Rate Seasonally Adj.)

[1] Except for the base year 1953, no year during which a recession was in process is included.

[2] About 32.0 million man-years of unemployment (true level) would have been consistent with maximum employment.

[3] Estimated as the difference between the officially reported civilian labor force and its likely size under conditions of maximum employment.

[4] In deriving these percentages, the civilian labor force is estimated as the officially reported civilian labor force plus concealed unemployment.

[5] Based upon sufficient annual rate of growth in G.N.P. to provide full use of growth in labor force, plant and productivity under conditions of maximum employment and production.

# AMOUNT BY WHICH TRUE LEVEL OF UNEMPLOYMENT HAS EXCEEDED LEVEL CONSISTENT WITH FULL EMPLOYMENT[1]/

### Millions of Workers

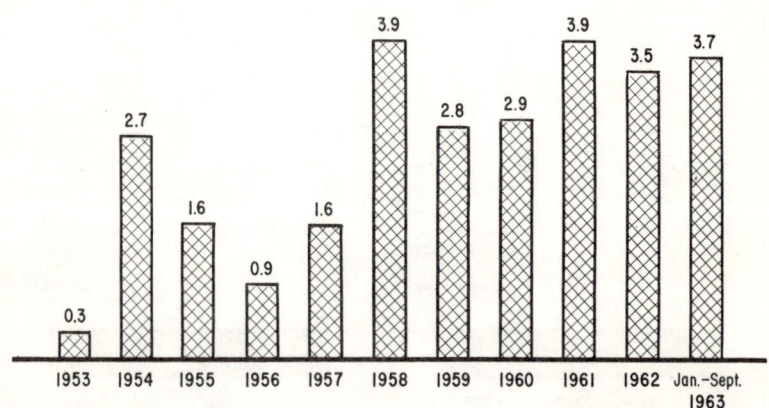

| | | | | | | | | | | |
|---|---|---|---|---|---|---|---|---|---|---|
| 0.3 | 2.7 | 1.6 | 0.9 | 1.6 | 3.9 | 2.8 | 2.9 | 3.9 | 3.5 | 3.7 |
| 1953 | 1954 | 1955 | 1956 | 1957 | 1958 | 1959 | 1960 | 1961 | 1962 | Jan.–Sept. 1963 (Seas. Adj.) |

## DURATION OF UNEMPLOYMENT

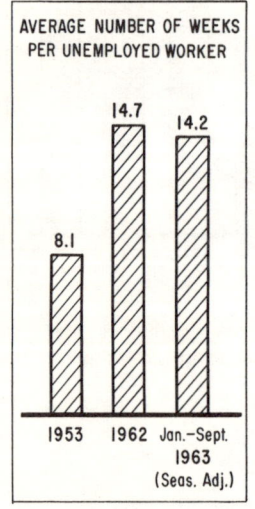

**AVERAGE NUMBER OF WEEKS PER UNEMPLOYED WORKER**

| 8.1 | 14.7 | 14.2 |
|---|---|---|
| 1953 | 1962 | Jan.–Sept. 1963 (Seas. Adj.) |

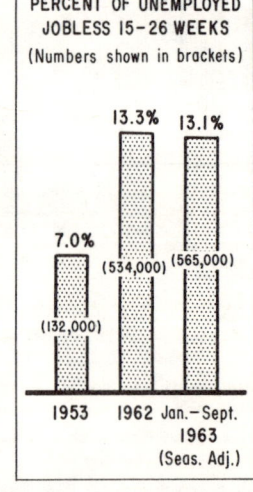

**PERCENT OF UNEMPLOYED JOBLESS 15-26 WEEKS**
(Numbers shown in brackets)

| 7.0% (132,000) | 13.3% (534,000) | 13.1% (565,000) |
|---|---|---|
| 1953 | 1962 | Jan.–Sept. 1963 (Seas. Adj.) |

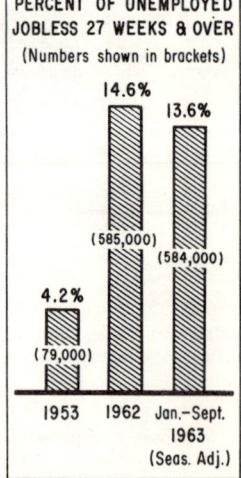

**PERCENT OF UNEMPLOYED JOBLESS 27 WEEKS & OVER**
(Numbers shown in brackets)

| 4.2% (79,000) | 14.6% (585,000) | 13.6% (584,000) |
|---|---|---|
| 1953 | 1962 | Jan.–Sept. 1963 (Seas. Adj.) |

1/ Full employment is here regarded as true level of unemployment equal to 4.5 percent of the civilian labor force, which equates with full-time recorded unemployment of 2.9 percent of civilian labor force.

# TOTAL OF THOSE UNEMPLOYED
## SHOWN BY CATEGORY, 1963 [1]

### (All Categories Add to 100 Percent)

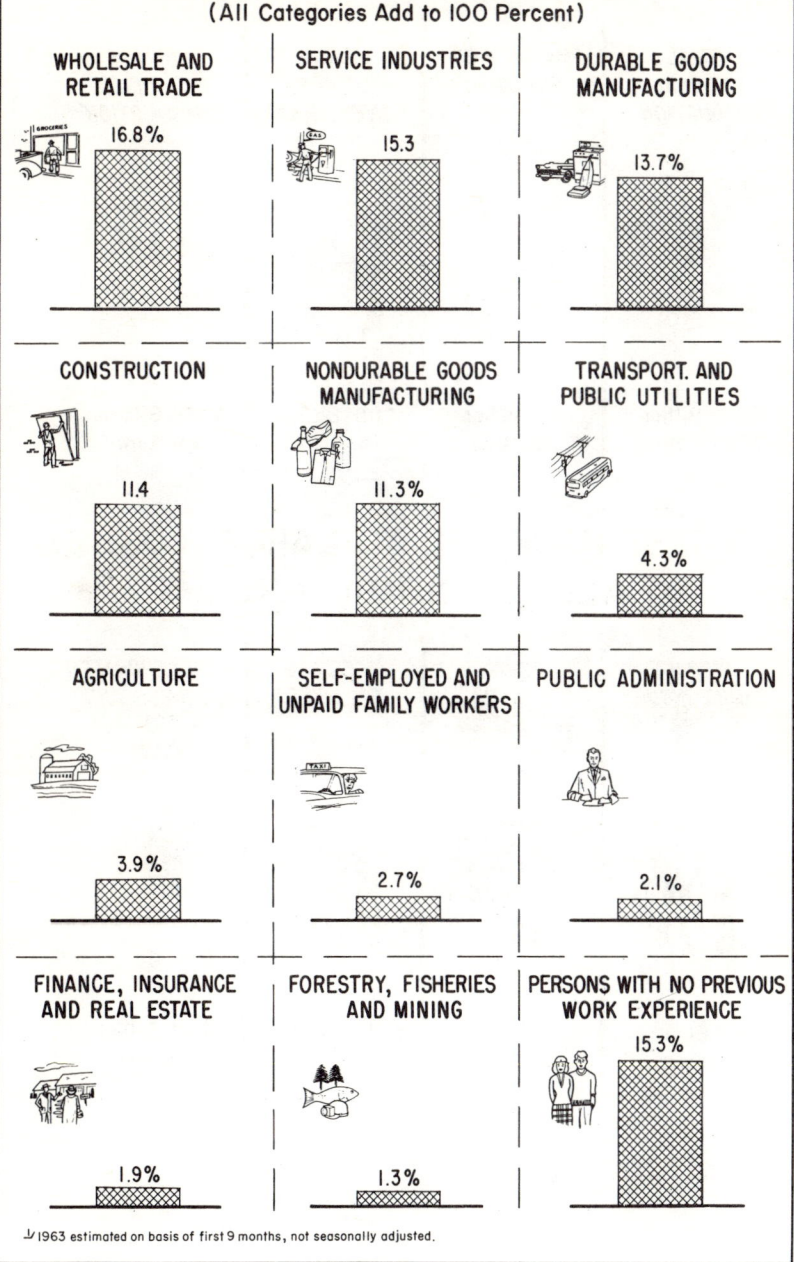

| WHOLESALE AND RETAIL TRADE | SERVICE INDUSTRIES | DURABLE GOODS MANUFACTURING |
| --- | --- | --- |
| 16.8% | 15.3 | 13.7% |

| CONSTRUCTION | NONDURABLE GOODS MANUFACTURING | TRANSPORT. AND PUBLIC UTILITIES |
| --- | --- | --- |
| 11.4 | 11.3% | 4.3% |

| AGRICULTURE | SELF-EMPLOYED AND UNPAID FAMILY WORKERS | PUBLIC ADMINISTRATION |
| --- | --- | --- |
| 3.9% | 2.7% | 2.1% |

| FINANCE, INSURANCE AND REAL ESTATE | FORESTRY, FISHERIES AND MINING | PERSONS WITH NO PREVIOUS WORK EXPERIENCE |
| --- | --- | --- |
| 1.9% | 1.3% | 15.3% |

[1] 1963 estimated on basis of first 9 months, not seasonally adjusted.

# LARGE NATIONAL ECONOMIC DEFICITS
## DURING PERIOD 1953–3rd QUARTER 1963

Dollar Items in 1962 Dollars

| TOTAL NATIONAL PRODUCTION (GNP) | MAN YEARS OF EMPLOYMENT | PRIVATE BUSINESS INVESTMENT (Incl. Net Foreign) | PRIVATE AND PUBLIC CONSUMPTION [1] |
|---|---|---|---|
|  |  |  |  |
| $475 Billion Too Low | 28 Million Too Low | $115 Billion Too Low | $360 Billion Too Low |

# ...THESE HAVE LED TO LARGE LOSSES
## TO ALL ECONOMIC GROUPS

| AVERAGE FAMILY INCOME (Multiple-Person Families) | FARM OPERATORS' NET INCOME | WAGES AND SALARIES | UNINCORPORATED BUSINESS AND PROFESSIONAL INCOME |
|---|---|---|---|
|  |  |  |  |
| $7,600 Too Low | $71 Billion Too Low | $313 Billion Too Low | $34 Billion Too Low |

[1] Includes personal consumption expenditures plus government (Federal, state, and local) expenditures ( 315 and 45 billions, respectively).

12

# III. Rising Unemployment
# Reflects Low Economic Growth

To avoid rising unemployment, our overall economic growth rate (growth in total national production, or GNP) must be large enough (a) to absorb the growth in the number of people able and willing to work, and (b) to prevent the rise in productivity or production per man-hour worked from cutting down the number of needed workers.

An average annual growth rate well in excess of 4 percent was needed during 1953-1963 to avoid this rising unemployment. But as shown by the chart on page 14, our actual economic growth rate averaged annually only 2.9 percent during these years.

This poor growth performance has followed a fairly regular pattern of recessions or absolute downturns in economic activity, recovery movements of insufficient speed and duration to bring us back to maximum prosperity, and periods of stagnation or abnormally low economic growth leading toward the next recession.

## The current "recovery" thus far is unsatisfactory

It is true that the current "recovery" movement offers prospect of lasting longer than the earlier "recovery" movements since 1953. But as the chart on page 14 shows, the swift upturn during the early stages of the current "recovery" was succeeded by a dwindling rate of economic growth in each successive twelve-month period from first quarter 1961-first quarter 1962 until the latest period shown on the chart, when the quickening was very slight. Indeed, from fourth quarter 1961 through third quarter 1963, the annual growth rate averaged below 4 percent. This should not be contrasted with the rate of 5 percent now required (in view of the new technology and automation) to maintain maximum employment and production *after* they have been achieved. This below 4 percent figure should be contrasted rather with the 8-9 percent annual growth rate (more than twice as high) required for at least two years to *lift us from where we are now* to maximum employment and production.

Although the "recovery" has by now been in process for almost three years, unemployment was higher during the first three quarters of 1963 seasonably adjusted than during 1962 (and rose still higher in November 1963), and the deficiency in total production was also larger. A patient, who has not recovered fully for an unusually long time, and even lost some ground, can hardly be comforted by the length of his 'recovery" movement.

The following chart rounds out this chapter.

# GROWTH RATES, U.S. ECONOMY, 1922-1963[1]

### Average Annual Rates Of Change In Gross National Product In Uniform 1962 Dollars

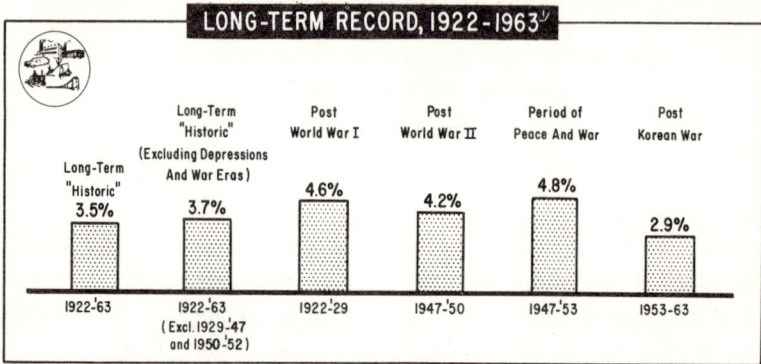

**LONG-TERM RECORD, 1922-1963[1]**

| Long-Term "Historic" 3.5% 1922-'63 | Long-Term "Historic" (Excluding Depressions And War Eras) 3.7% 1922-'63 (Excl. 1929-'47 and 1950-'52) | Post World War I 4.6% 1922-'29 | Post World War II 4.2% 1947-'50 | Period of Peace And War 4.8% 1947-'53 | Post Korean War 2.9% 1953-63 |

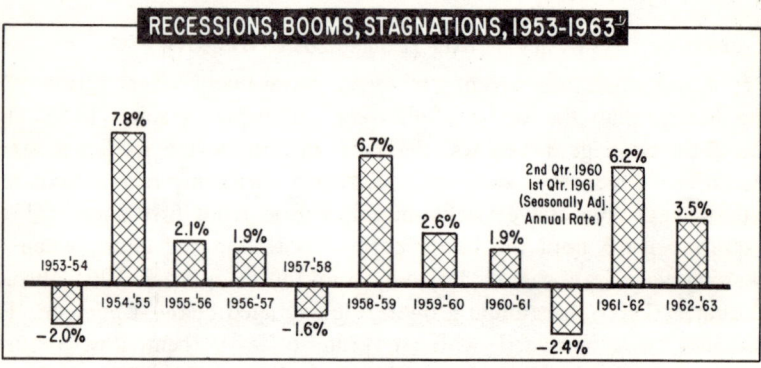

**RECESSIONS, BOOMS, STAGNATIONS, 1953-1963[1]**

- 1953-'54: -2.0%
- 1954-'55: 7.8%
- 1955-'56: 2.1%
- 1956-'57: 1.9%
- 1957-'58: -1.6%
- 1958-'59: 6.7%
- 1959-'60: 2.6%
- 1960-'61: 1.9%
- 2nd Qtr. 1960 - 1st Qtr. 1961 (Seasonally Adj. Annual Rate): -2.4%
- 1961-'62: 6.2%
- 1962-'63: 3.5%

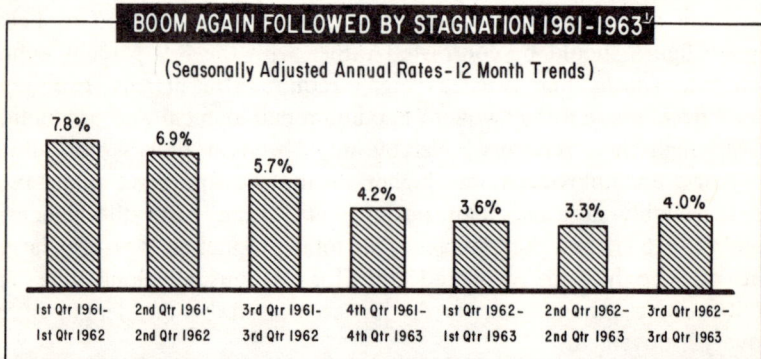

**BOOM AGAIN FOLLOWED BY STAGNATION 1961-1963[1]**

(Seasonally Adjusted Annual Rates-12 Month Trends)

| 1st Qtr 1961- 1st Qtr 1962 | 2nd Qtr 1961- 2nd Qtr 1962 | 3rd Qtr 1961- 3rd Qtr 1962 | 4th Qtr 1961- 4th Qtr 1963 | 1st Qtr 1962- 1st Qtr 1963 | 2nd Qtr 1962- 2nd Qtr 1963 | 3rd Qtr 1962- 3rd Qtr 1963 |
| 7.8% | 6.9% | 5.7% | 4.2% | 3.6% | 3.3% | 4.0% |

[1] 1963 estimated on basis of first three quarters.

14

# IV. The Main Trouble:
# Deficiency In Demand Or Purchasing Power

When unemployment and idle plant have been rising chronically for a decade, the reason must be insufficient demand (or purchasing power translated into actual purchases) to call forth full use of our productive resources. The total of this demand for goods and services equals total national production (GNP). When this total demand is too low, production is too low and unemployment too high.

This total demand has three large components: private consumer expenditures, public outlays for goods and services at all levels of government, and gross private investment. The size and causes of the deficiencies in these three types of demand explain the poor economic performance.

## The dominating deficiency is private consumer spending

Private consumer spending is the largest component in total demand, usually above three-fifths of the total. And during the past decade, as shown by the chart on page 17, consumer spending has fallen very far short of the needed rate of growth in almost every year. The average annual consumer spending deficiency during 1953-1963 has been about 30 billion dollars, or two-thirds of the average annual deficiency of about 45 billion in total demand. In 1963 alone, the estimated consumer spending deficiency of 59 billion loomed large in the total demand deficiency of 76.5 billion.

Contrary to popular belief, inadequate consumer spending is not due to decisions by American families to spend too small a part of their incomes and save too large a part. As shown by the chart on page 18, the main cause is inadequate consumer income. Another factor is the unsatisfactory distribution of income: the 8 percent of all American families with incomes of $15,000 and over receive about 24 percent of total family income, while the 23 percent of all families with incomes below $4,000 receive only about 7 percent. This aggravates the consumer spending deficiency, because those at the top of the income structure save and seek to invest relatively larger portions of their incomes, while those lower down spend for immediate consumption as much as they earn after paying taxes, or spend more and go into debt.

## The deficiency in public outlays for goods and services

As the chart on page 17 shows, the period 1953-1963 registered

15

an average annual deficiency of 4.2 billion dollars in public outlays for goods and services. This has reflected mainly an excessively tight Federal Budget. The States and localities (unlike the Federal Government) have increased their spending and debts at an exceedingly rapid pace, despite much more limited power to tax and borrow than the Federal Government.

The shrinkage in the Federal Budget on a per capita basis relative to population is shown by the chart on page 19. Comparing fiscal 1964 with fiscal 1954, sharp reductions in per capita outlays for national security and international purposes (including space research and technology) have not been sufficiently compensated for by increased spending for seriously neglected domestic purposes. The Budget has also shrunk, measured as a percentage of total national production (or demand).

The hurtful effects far exceed these figures. Due to its "multiplier" effect upon private outlays, every dollar of Federal spending adds about three to four dollars to total demand.

## The problem of business investment

For the period 1953-1963 *as a whole,* business investment also was deficient. But as the chart on page 20 shows, business investment in plant and equipment grew more than three times as fast as demand for ultimate goods and services (in the form of private consumer spending and public outlays) during the upturn period preceding the 1957-1958 recession: during the upturn period preceding the 1960-1961 recession, investment grew more than 4½ times as fast; and during the upturn period from first quarter 1961 to third quarter 1963, it grew more than 1½ times as fast.

These excessive and nonsustainable investment spurts have not been held back by any excessive tax burden; they have been supported by more than ample after-tax profits, retained earnings, and other sources of financing. The chart on page 21 indicates, through second quarter 1963, the tendency of key profits after taxes to reach new peaks levels during such upturn periods, despite very large unused plant capacities.

The periodic sharp downturns in investment in plant and equipment have come only when the persistent deficiencies in private consumer spending and public outlays resulted in very large idle plant capacities. A better balance between (a) investment and (b) private consumer spending and public outlays would stabilize and thus enlarge long-term economic growth. This would help investment and profits in the long run.

The following five charts amplify this discussion.

16

# DEFICIENT RATE OF GROWTH IN PRIVATE CONSUMER SPENDING, 1953 - 1963[1/]

### Rates of Change in 1962 Dollars

Needed Rate of Growth    Actual Rate of Growth

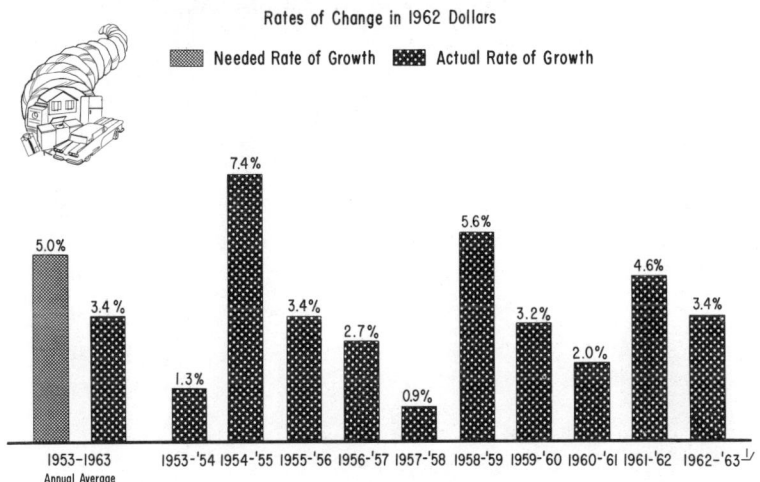

# THE PRIVATE CONSUMPTION DEFICITS DOMINATE THE DEFICITS IN THE TOTAL ECONOMY

### Billions of 1962 Dollars

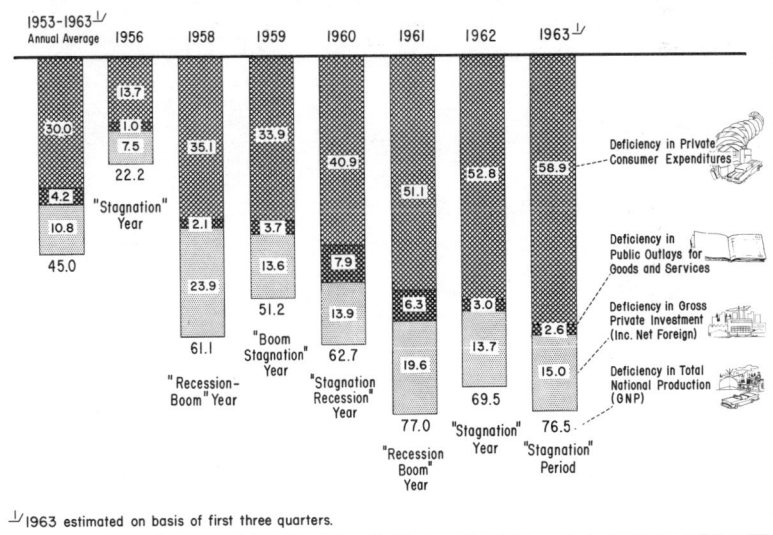

[1/] 1963 estimated on basis of first three quarters.

# LOW GROWTH IN PRIVATE CONSUMPTION REFLECTS EVEN LOWER GROWTH IN INCOMES IN MOST OF RECENT YEARS

Rates of Change in 1962 Dollars

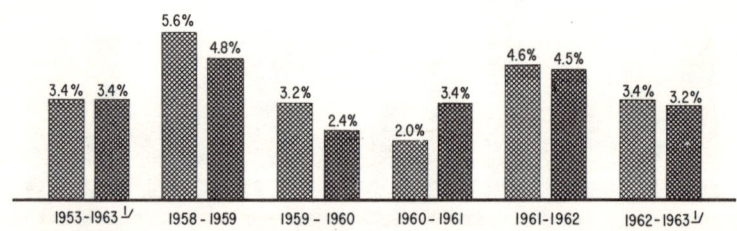

# THE PRIVATE CONSUMPTION DEFICIENCY OF $ 315 BILLION, 1953 - 3rd Qtr. 1963 REFLECTED A $ 410 BILLION INCOME DEFICIENCY

Billions of 1962 Dollars

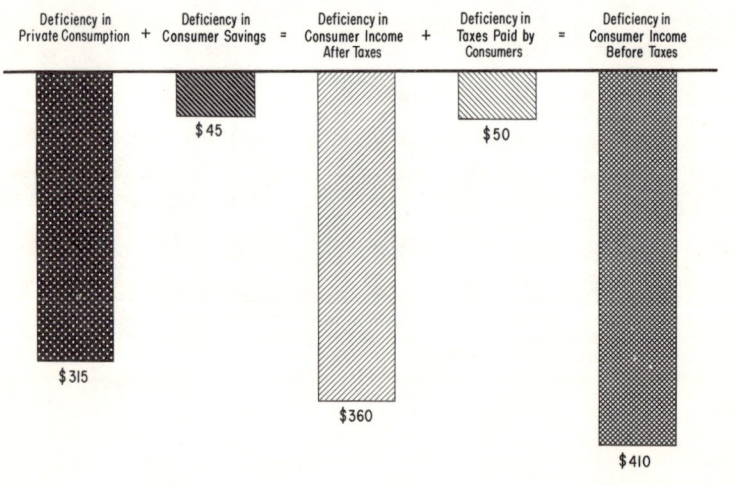

1/1963 estimated on basis of first three quarters.

18

# FEDERAL BUDGET HAS SHRUNK RELATIVE TO SIZE OF ECONOMY AND NEEDS, 1954-'64

### Fiscal Years

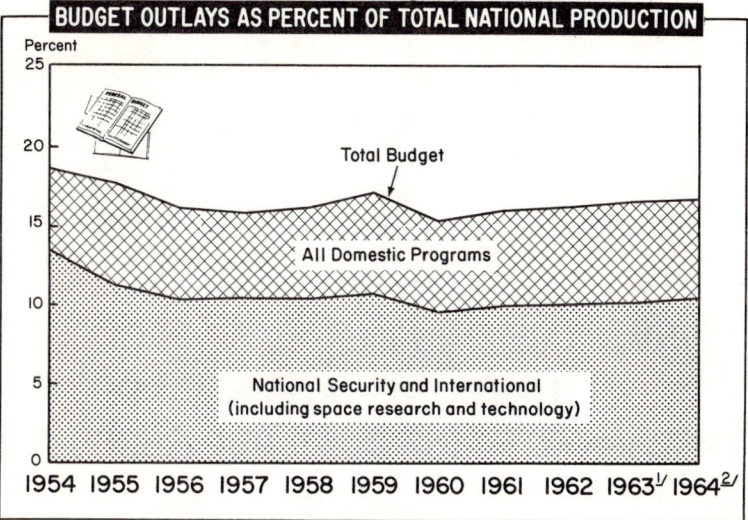

**BUDGET OUTLAYS AS PERCENT OF TOTAL NATIONAL PRODUCTION**

Percent

Total Budget

All Domestic Programs

National Security and International
(including space research and technology)

1954 1955 1956 1957 1958 1959 1960 1961 1962 1963[1] 1964[2]

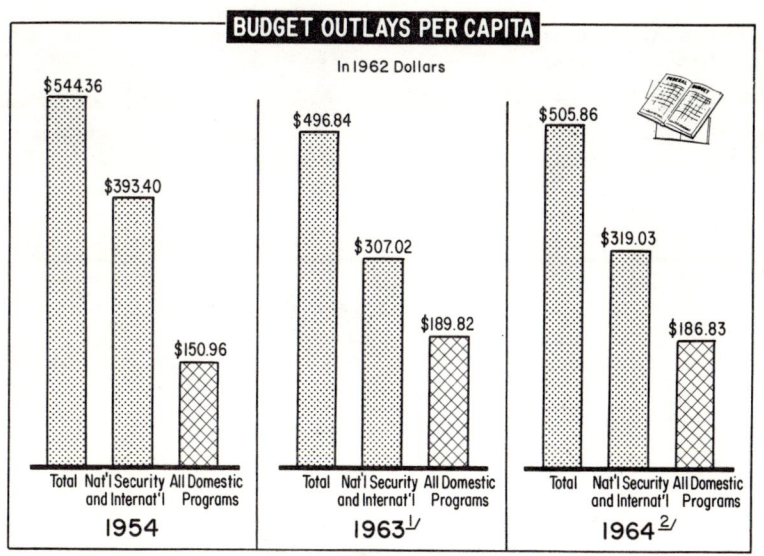

**BUDGET OUTLAYS PER CAPITA**

In 1962 Dollars

$544.36

$393.40

$150.96

| Total | Nat'l Security and Internat'l | All Domestic Programs |
|---|---|---|

**1954**

$496.84

$307.02

$189.82

| Total | Nat'l Security and Internat'l | All Domestic Programs |
|---|---|---|

**1963[1]**

$505.86

$319.03

$186.83

| Total | Nat'l Security and Internat'l | All Domestic Programs |
|---|---|---|

**1964[2]**

[1] Preliminary. G.N.P. estimated at $565 billion, CEP.

[2] Administration's proposed Budget as of Jan. 17, 1963. G.N.P. estimated at $588 billion, CEP.

19

# INVESTMENT IN PLANT AND EQUIPMENT WAS DEFICIENT –1953–1963 AS A WHOLE [1]

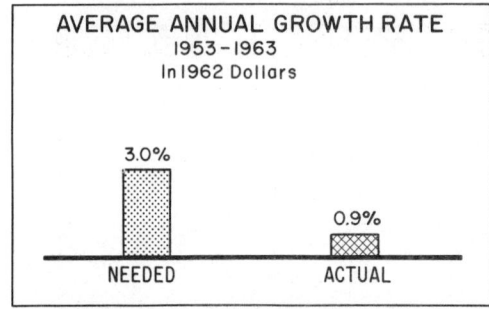

AVERAGE ANNUAL GROWTH RATE
1953–1963
In 1962 Dollars

3.0%
NEEDED

0.9%
ACTUAL

AVERAGE ANNUAL DEFICIENCY
1953–1963
Billions of 1962 Dollars

$5.3

# BUT INVESTMENT IN MEANS OF PRODUCTION AT TIMES OUTRAN DEMAND; HENCE INVESTMENT CUTS AND RECESSIONS

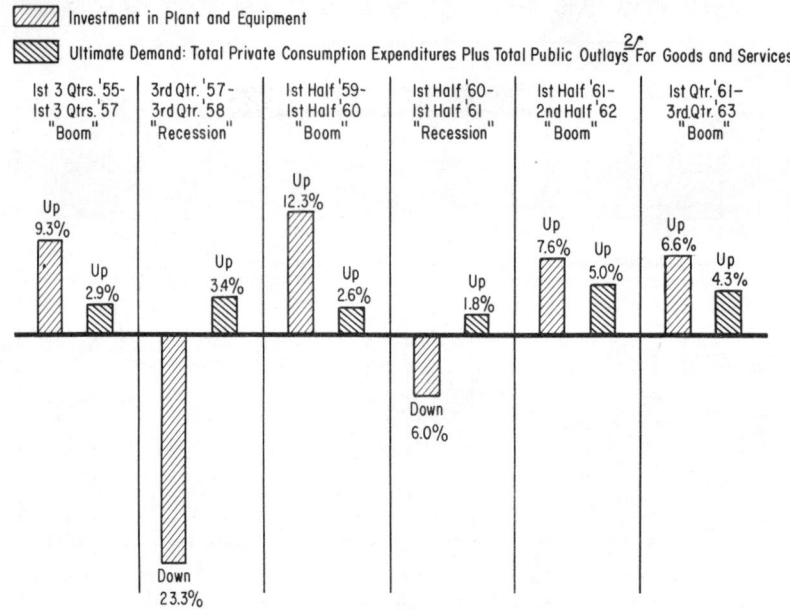

Investment in Plant and Equipment

Ultimate Demand: Total Private Consumption Expenditures Plus Total Public Outlays [2] For Goods and Services

| 1st 3 Qtrs. '55– 1st 3 Qtrs. '57 "Boom" | 3rd Qtr. '57– 3rd Qtr. '58 "Recession" | 1st Half '59– 1st Half '60 "Boom" | 1st Half '60– 1st Half '61 "Recession" | 1st Half '61– 2nd Half '62 "Boom" | 1st Qtr. '61– 3rd Qtr. '63 "Boom" |
|---|---|---|---|---|---|
| Up 9.3% / Up 2.9% | Down 23.3% / Up 3.4% | Up 12.3% / Up 2.6% | Down 6.0% / Up 1.8% | Up 7.6% / Up 5.0% | Up 6.6% / Up 4.3% |

AVERAGE ANNUAL RATES OF CHANGE
In 1962 Dollars

[1] 1963 estimated on basis of first three quarters.
[2] Federal, State and local.

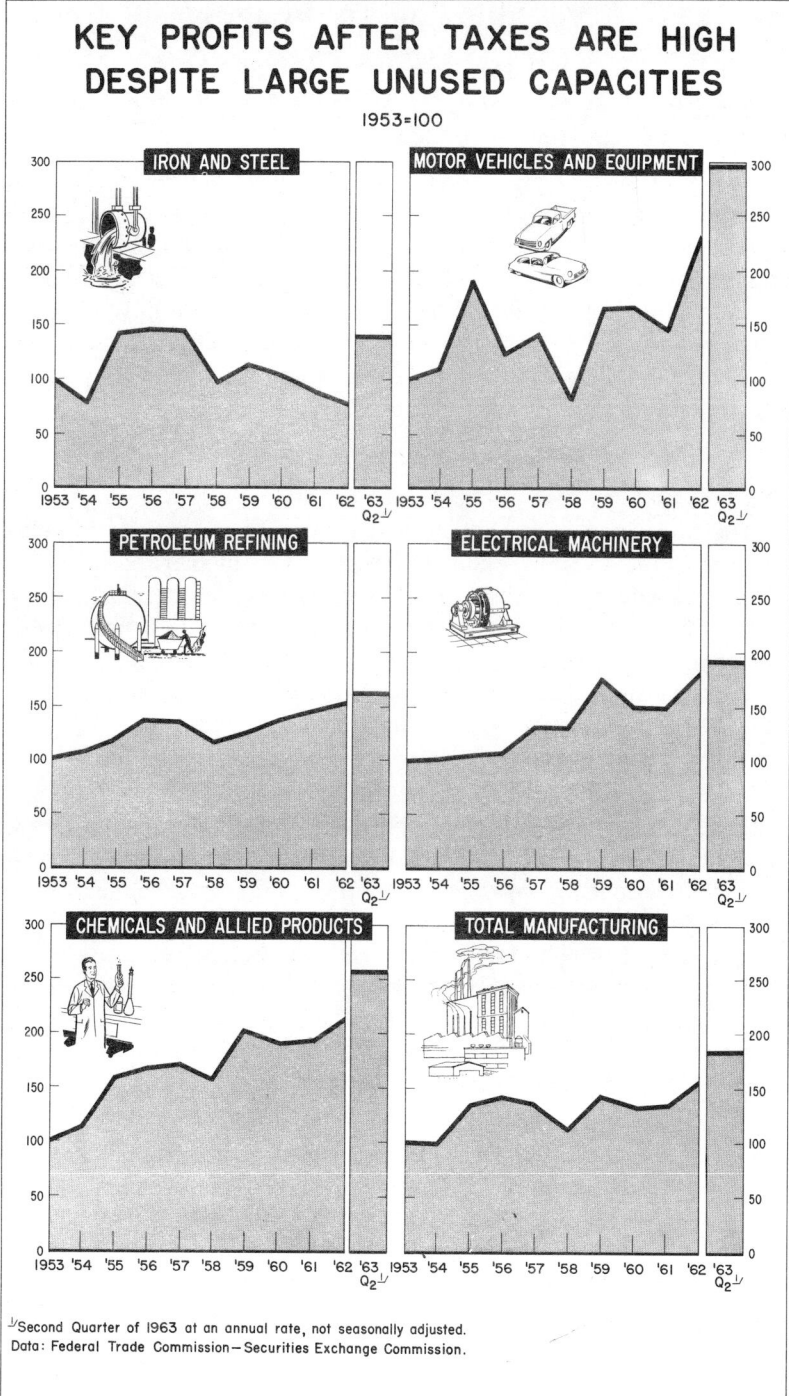

# KEY PROFITS AFTER TAXES ARE HIGH
# DESPITE LARGE UNUSED CAPACITIES

1953=100

⅃/Second Quarter of 1963 at an annual rate, not seasonally adjusted.
Data: Federal Trade Commission—Securities Exchange Commission.

21

# V. The Lag In Wages
# And The Wage Deficiency

## The huge size of the wage deficiency

The previous chapter noted that the huge deficiency in private consumption has stemmed from an even larger deficiency in personal income before taxes. And as shown by the chart on page 26, an average annual deficiency of 29.8 billion dollars in wages and salaries was the major part of an average annual deficiency of 39.1 billion in total personal incomes before taxes during 1953-1963. During 1963 alone, a deficiency of 56.4 billion dollars in wages and salaries was more than 77 percent of the total personal income deficiency of 72.9 billion.

## Wages have lagged far behind productivity gains

The wage inadequacy has been due in large measure to insufficient employment and excessive unemployment. But another basic cause of the wage inadequacy, which in turn has led to much of the unemployment, has been the serious lag in hourly wage rate increases behind gains in productivity or output per man-hour worked. Yet there has been widespread propaganda that wage rate increases have been advancing more rapidly than productivity gains, thus inflating prices and hurting our competitive position overseas.

In the railroad industry, as shown by the chart on page 27, where there has been so much talk about "featherbedding" and other allegations of undesirable labor practices, real hourly earnings for all railroad workers during the period 1957-1962 increased at an average annual rate of only 2.3 percent. But productivity increased at an average annual rate of 5.5 percent. During the same five years, in iron and steel, another industry where the hue and cry about "excessive" wage increases has been continuous, the average annual rate of increase in the real hourly earnings of production workers was only 2.6 percent. But the average annual increase in their productivity was 3.4 percent.

It is sometimes argued that the rate of increases in real hourly earnings in specific industries should approximate the productivity trends for the private economy as a whole, rather than the productivity trends in these specific industries. But the same chart shows that the above-cited 2.6 percent average annual rate of increases in the real hourly wages of production workers in iron and steel, 1957-1962, was much lower than the 3.4 percent average annual rate of productivity gain for all manufacturing wage and salary employment, and very slightly lower than the 2.7 percent average annual rate of productivity gain for all private non-farm wage and salary employment. And the above-cited 2.3 percent average annual rate of increase in the real hourly wages and salaries of railroad workers was lower by an even wider margin than the productivity gains for private nonfarm employment or for manufacturing.

Without reference to particular industries: In the whole private non-farm economy, while productivity rose at the above-cited average annual rate of 2.7 percent, real hourly earnings of all wage and salary employees rose at an average annual rate of only 2.1 percent. In manu-facturing, while productivity rose at the above-cited average annual rate of 3.4 percent, real hourly earnings of all wage and salary employees rose at an average annual rate of only 1.8 percent.

## Appropriate standards for wage rate increases

In industries with unusually low productivity gains, it is obvious that the wage rate increases will lag behind the average productivity gain for the whole private economy. Thus, if wage rate increases in industries registering very high productivity gains are held down to the average productivity gain for the whole private economy, in accord with the prevalent theory, the average rate of wage increases for the whole private economy will fall far below the average productivity gain.

There might be some merit in the prevalent theory, if the excessively high profits resulting from wage rate gains lagging behind productivity gains in the very high productivity industries were taxed away, and the proceeds used for public programs to increase the incomes and living standards of low-income workers; or if prices were lowered substantially in these high productivity industries, thus passing on to the consumer the gap between the high productivity gains and the lower rate of wage increases. But none of this has happened, and no programs to help it happen are in the offing.

This explains the damaging effects of the Government's so-called wage guidelines, suggesting that wage rate increases in particular industries be lim-ited to the average gain in productivity for the whole private economy. These guidelines have caused many workers to accept wage rate gains, not only below their own productivity gains, but far below the productivity gain for the whole private economy. The Government should be warning against the injurious effects of the lag in wage rate increases behind productivity gains.

## The distinction between actual productivity trends and true trends in our productivity-potential

Even if wage rate gains during the past five years had been kept in line with actual productivity gains (which did not happen), the wage rate gains would have been too low from the viewpoint of our economic needs. This is because wage rate gains, in order to fulfil their role in the expansion of consumption, should equal the productivity-potential gains embodied in the advance of technology and automation. These productivity-potential gains, during recent years, have been much higher than the actual productivity gains, as usually computed, because the latter have been repressed by the inefficiencies resulting from large unused plant and low economic growth. To adjust wage rate changes to these repressed productivity gains holds the expansion of consumption below the rate required to use our available resources fully. This compounds the evils of economic slack, instead of helping to overcome them.

The difference between actual productivity gains and true productivity-potential gains might well be called concealed productivity gains, just as concealed unemployment represents the difference between actual growth in the civilian labor force and what this growth would be if the economy were functioning better. Insofar as even the inadequate current economic upturn has called forth some of the concealed productivity-potential and the concealed labor force potential, there has not been nearly as much reduction of recorded unemployment as had been expected by those who have been underestimating the pace of the new technology and automation. In consequence of these underestimates, the overall economic growth goals which the Government has set are much too low, even if achieved, to restore maximum employment.

## Tendency toward acceleration of productivity gains

The chart on page 28 shows the long-range trend toward an accelerating rate of actual productivity gains. During the period 1955-1961, the extremely large economic slack and the two recessions caused the actual rate of productivity gains to fall off. But in 1961-1962, the actual rate of productivity gains moved closer to the productivity-potential gains, thus bringing to the surface the long-range trend toward acceleration. It is to these productivity-potential trends that wage rates should be adjusted. Moreover, the increasing productivity of capital—the fact that each dollar of added invesment in plant and equipment adds more to output capabilities than it did many years ago—is another reason for drastic reconsideration ·of the whole problem of wage rate increases as these bear upon a

24

satisfactory relationship between expansion of consumer incomes and expansion of our ability to produce.

## Comparative trends in profits, investment, and wages

It may also be argued that adjusting wage rate gains to the real trends in the productivity-potential, rather than to the repressed actual productivity gains resulting from economic slack, would impose an excessive squeeze or pinch upon profits and investment. But as shown by the chart on page 29, profits after taxes and especially investment in plant and equipment before the 1957-1958 recession far outran increases in wage rates. The chart on page 30 shows that, despite profits reduced from excessively high levels, investment in plant and equipment before the 1960-1961 recession again outran wage rate increases by enormous margins. And the chart on page 31 shows, for the period from first quarter 1961 to second quarter 1963, profits after taxes tremendously outrunning wage rate increases. The relationship of investment increases to wage rate increases during this most recent period is more irregular. This is because, even with profits breaking all previous records, a high rate of investment advance is unlikely when there is so much idle plant capacity, and when the demand for ultimate products in the form of private consumption and public outlays is so seriously deficient.

## Relationship between wage rate increases and fringe benefits

The foregoing discussion of wage rate increases does not take account of fringe benefits. It should not, although improved fringe benefits are desirable. It is an expanding volume of wages, and not the volume of fringe benefits, which adds mainly to the expansion of consumption— and this expansion of consumption is the great need. The cost of fringe benefits does have a bearing upon the adequacy of profit margins. But as already shown in detail, these profit margins have been more than rewarding.

The following six charts illustrate this chapter.

25

# DEFICIENT RATE OF GROWTH IN WAGES AND SALARIES, 1953– 1963 ⅃

### Rates of Change in 1962 Dollars

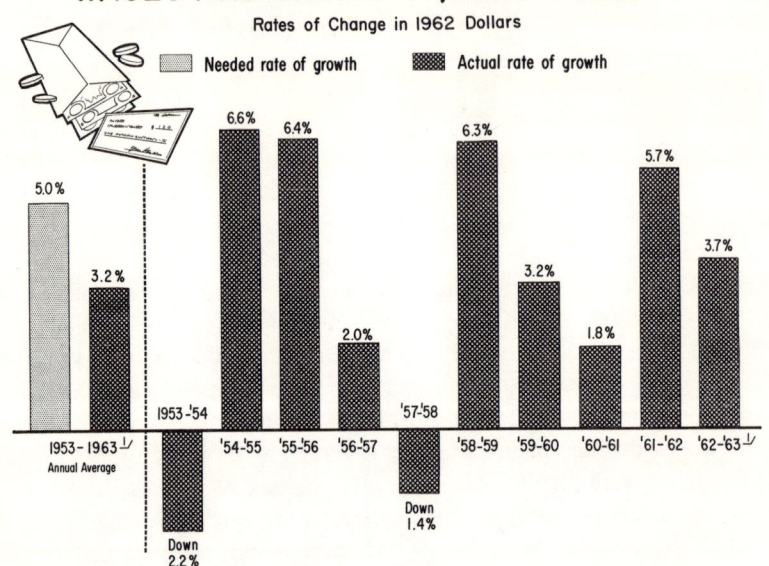

□ Needed rate of growth    ▨ Actual rate of growth

# DEFICIENCIES IN WAGES AND SALARIES ARE LARGE SHARE OF DEFICIENCIES IN TOTAL CONSUMER INCOMES BEFORE TAXES

### Billions of 1962 Dollars

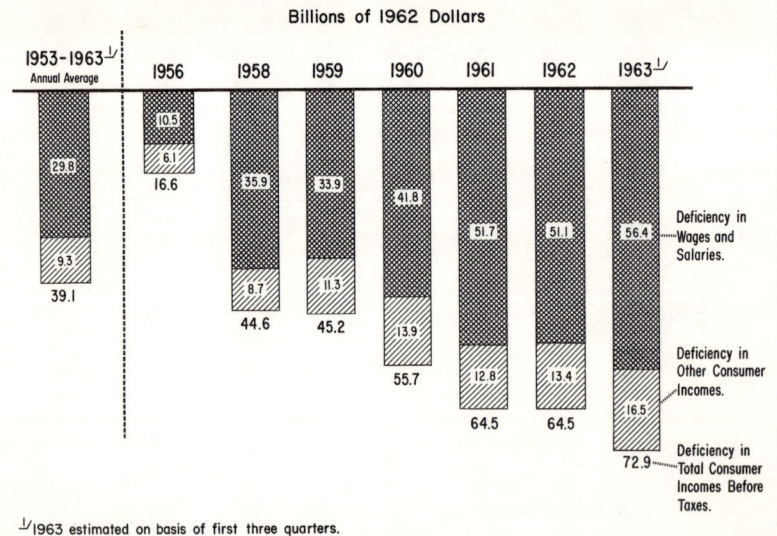

⅃1963 estimated on basis of first three quarters.

# COMPARATIVE TRENDS IN PRODUCTIVITY AND REAL HOURLY EARNINGS, 1957-1962

### Average Annual Rates of Change

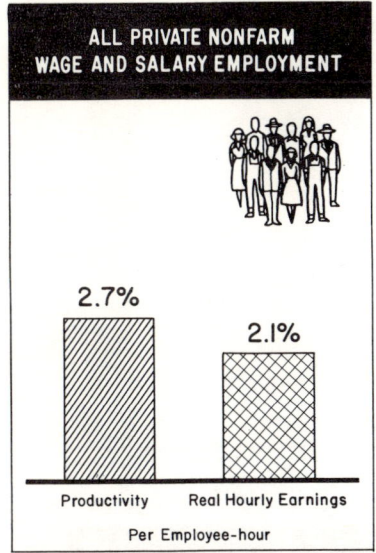

**ALL PRIVATE NONFARM WAGE AND SALARY EMPLOYMENT**

2.7%
Productivity

2.1%
Real Hourly Earnings

Per Employee-hour

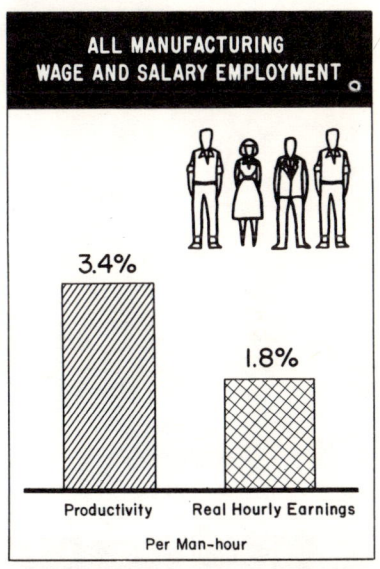

**ALL MANUFACTURING WAGE AND SALARY EMPLOYMENT**

3.4%
Productivity

1.8%
Real Hourly Earnings

Per Man-hour

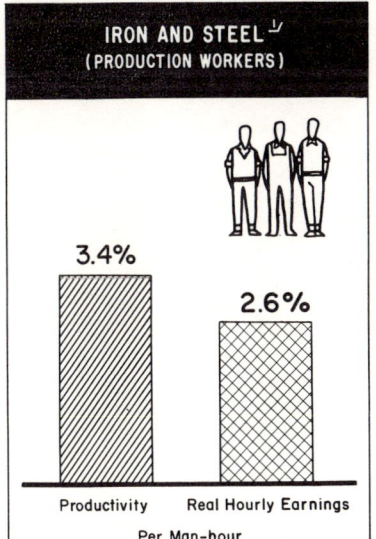

**IRON AND STEEL** [1] **(PRODUCTION WORKERS)**

3.4%
Productivity

2.6%
Real Hourly Earnings

Per Man-hour

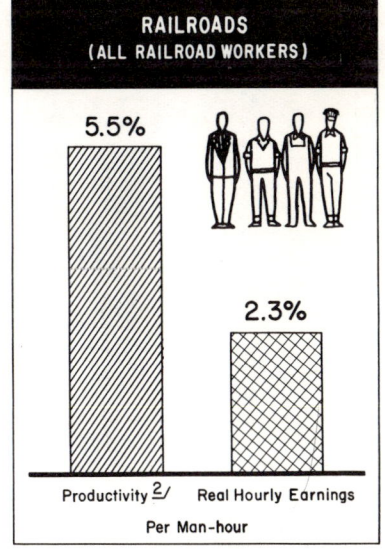

**RAILROADS (ALL RAILROAD WORKERS)**

5.5%
Productivity [2]

2.3%
Real Hourly Earnings

Per Man-hour

[1] Estimated by United Steelworkers of America.
[2] Productivity based on trends in traffic units per man-hour as reported by I.C.C.

Basic data: U.S. Dept. of Labor (except as noted)

27

# TRENDS IN PRODUCTIVITY FOR THE ENTIRE PRIVATE ECONOMY–1910–1962

### Average Annual Rate of Growth in Output per Man-hour for the Entire Private Economy

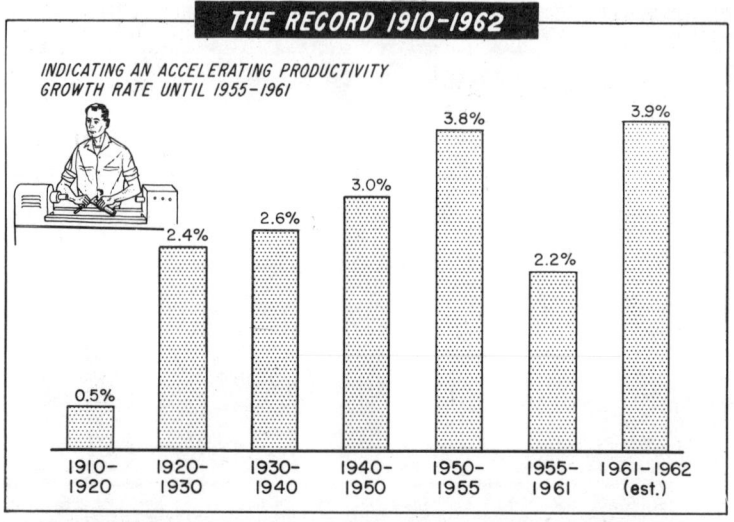

**THE RECORD 1910–1962**

INDICATING AN ACCELERATING PRODUCTIVITY
GROWTH RATE UNTIL 1955–1961

| | | | | | | |
|---|---|---|---|---|---|---|
| 0.5% | 2.4% | 2.6% | 3.0% | 3.8% | 2.2% | 3.9% |
| 1910–1920 | 1920–1930 | 1930–1940 | 1940–1950 | 1950–1955 | 1955–1961 | 1961–1962 (est.) |

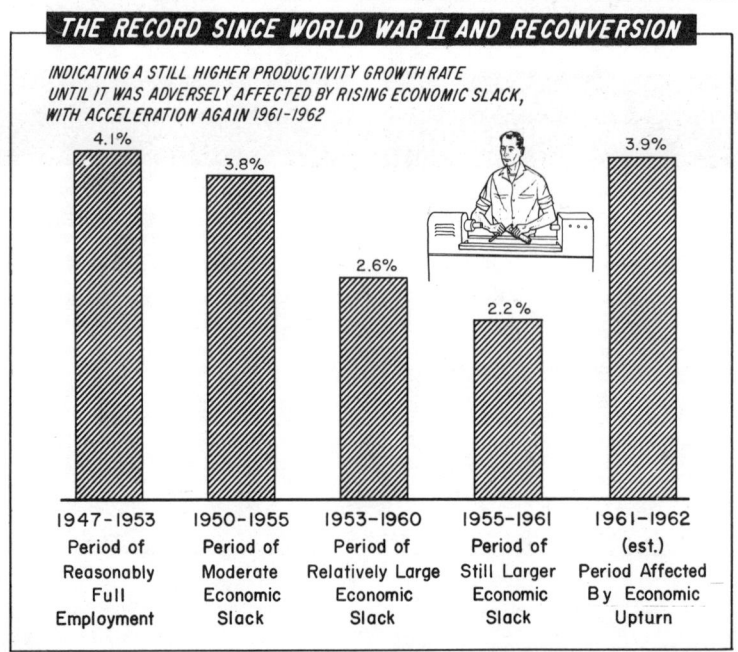

**THE RECORD SINCE WORLD WAR II AND RECONVERSION**

INDICATING A STILL HIGHER PRODUCTIVITY GROWTH RATE
UNTIL IT WAS ADVERSELY AFFECTED BY RISING ECONOMIC SLACK,
WITH ACCELERATION AGAIN 1961–1962

| 4.1% | 3.8% | 2.6% | 2.2% | 3.9% |
|---|---|---|---|---|
| 1947–1953 Period of Reasonably Full Employment | 1950–1955 Period of Moderate Economic Slack | 1953–1960 Period of Relatively Large Economic Slack | 1955–1961 Period of Still Larger Economic Slack | 1961–1962 (est.) Period Affected By Economic Upturn |

Note: Based on U.S. Department of Labor estimates relating to man-hours worked, based on labor force data.

# BEFORE THE 1957-1958 RECESSION, PROFITS AND INVESTMENT OUTRAN WAGES-BASIC TO CONSUMPTION

First Three Quarters 1955 — First Three Quarters 1957

▨ Profits after Taxes;[1]   ▨ Investment in Plant and Equipment;[2]   ▢ Wage Rates [3]

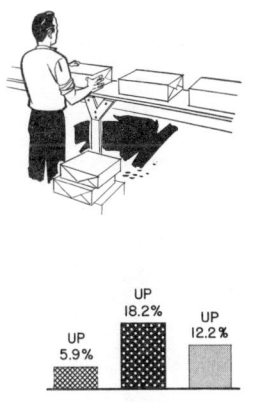

UP 5.9%
UP 18.2%
UP 12.2%

**PROCESSED FOODS and KINDRED PRODUCTS**

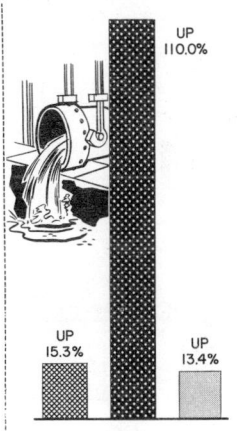

UP 110.0%
UP 15.3%
UP 13.4%

**IRON and STEEL**

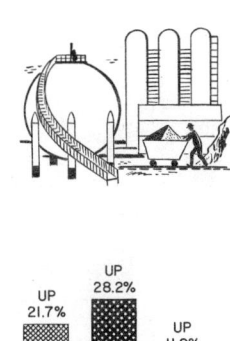

UP 21.7%
UP 28.2%
UP 11.9%

**PETROLEUM and COAL PRODUCTS**

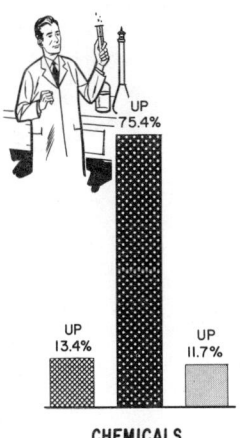

UP 75.4%
UP 13.4%
UP 11.7%

**CHEMICALS and ALLIED PRODUCTS**

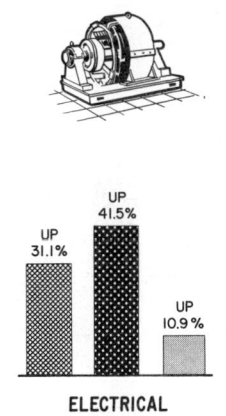

UP 31.1%
UP 41.5%
UP 10.9%

**ELECTRICAL MACHINERY**

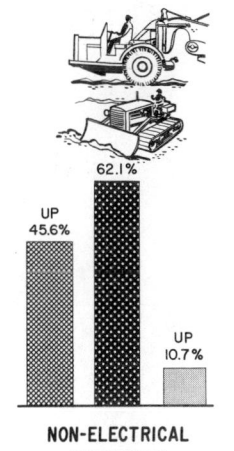

62.1%
UP 45.6%
UP 10.7%

**NON-ELECTRICAL MACHINERY**

[1] Data: Securities and Exchange Commission.
[2] Data: U.S. Dept. of Commerce and Securities and Exchange Commission.
[3] Average hourly earnings of production workers. Data: U.S. Dept. of Labor.

# BEFORE THE 1960-1961 RECESSION, DESPITE REDUCED PROFITS, INVESTMENT OUTRAN WAGES—BASIC TO CONSUMPTION

First Half 1959 – First Half 1960

▓ Profits after Taxes [1]      ▓ Investment in Plant and Equipment [2]      ▓ Wage Rates [3]

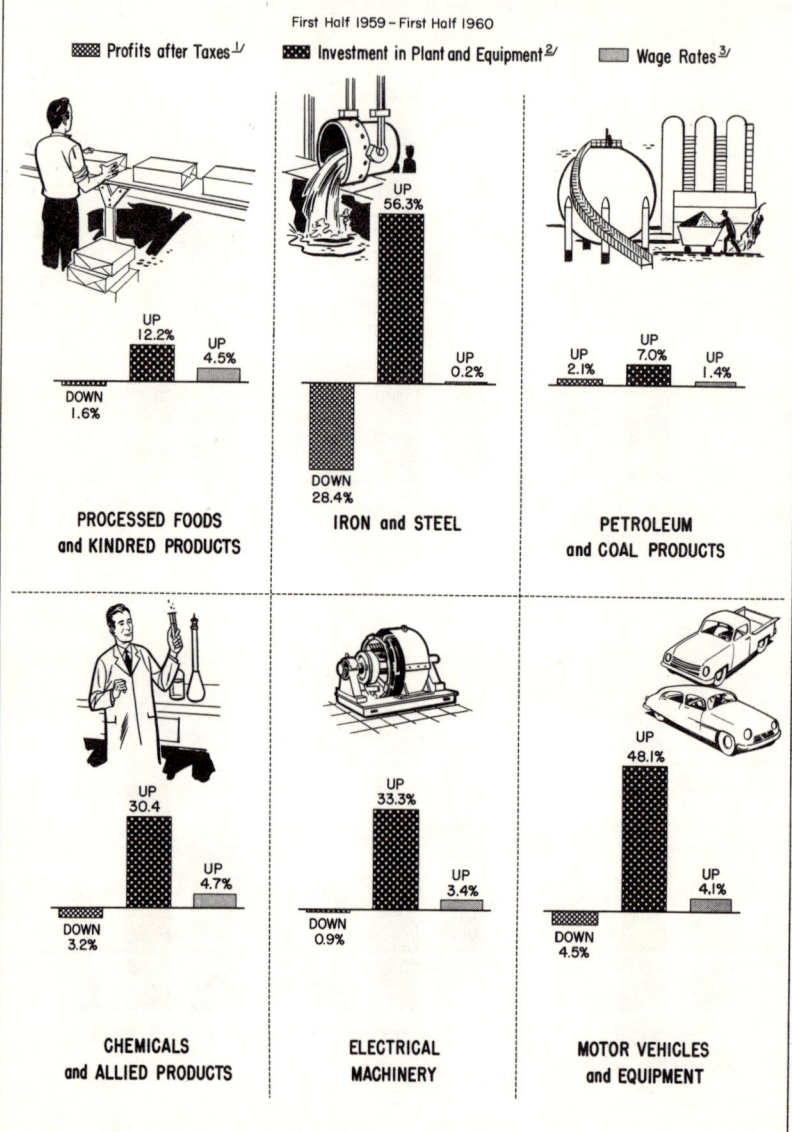

**PROCESSED FOODS and KINDRED PRODUCTS**
UP 12.2%
UP 4.5%
DOWN 1.6%

**IRON and STEEL**
UP 56.3%
UP 0.2%
DOWN 28.4%

**PETROLEUM and COAL PRODUCTS**
UP 2.1%
UP 7.0%
UP 1.4%

**CHEMICALS and ALLIED PRODUCTS**
UP 30.4
UP 4.7%
DOWN 3.2%

**ELECTRICAL MACHINERY**
UP 33.3%
UP 3.4%
DOWN 0.9%

**MOTOR VEHICLES and EQUIPMENT**
UP 48.1%
UP 4.1%
DOWN 4.5%

[1] Data: Securities and Exchange Commission
[2] Data: U. S. Dept. of Commerce and Securities and Exchange Commission
[3] Average hourly earnings of production workers. Data: Dept. of Labor

# DURING CURRENT ECONOMIC UPTURN, PROFITS, AND INVESTMENT IN SOME CASES, OUTRUN WAGES-BASIC TO CONSUMPTION

1st Quarter 1961 - 2nd Quarter 1963

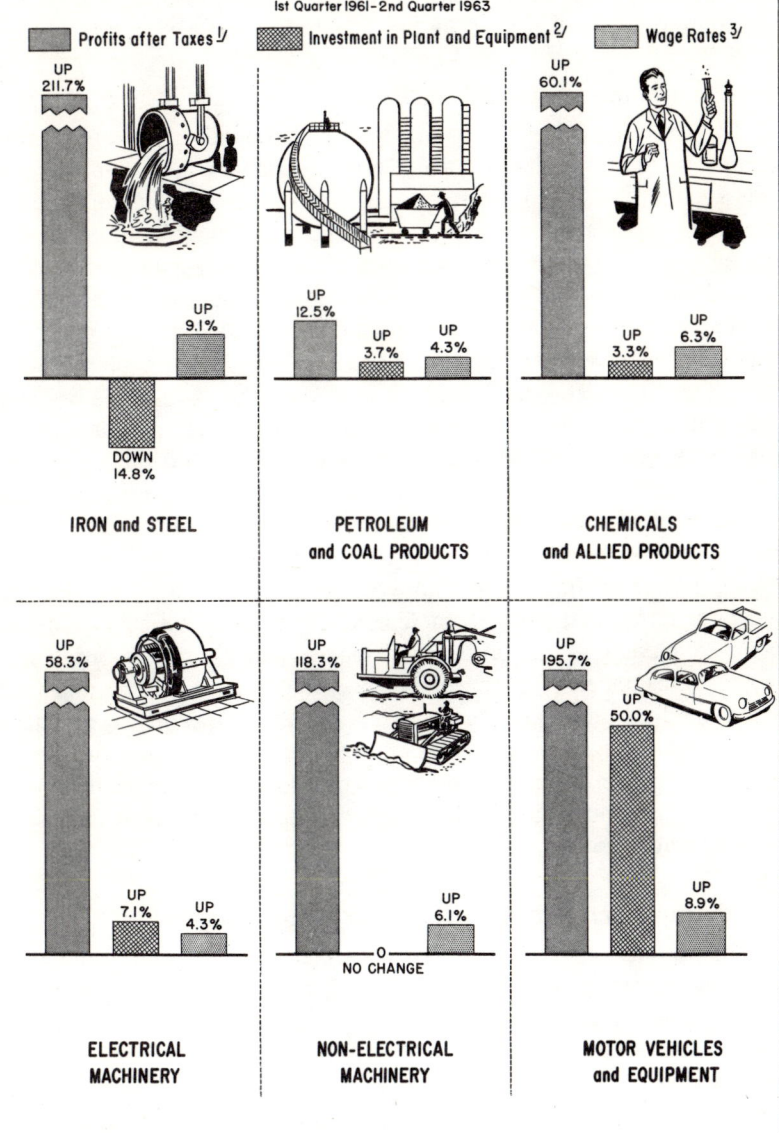

Profits after Taxes [1]  Investment in Plant and Equipment [2]  Wage Rates [3]

**IRON and STEEL**
UP 211.7%
UP 9.1%
DOWN 14.8%

**PETROLEUM and COAL PRODUCTS**
UP 12.5%
UP 3.7%
UP 4.3%

**CHEMICALS and ALLIED PRODUCTS**
UP 60.1%
UP 3.3%
UP 6.3%

**ELECTRICAL MACHINERY**
UP 58.3%
UP 7.1%
UP 4.3%

**NON-ELECTRICAL MACHINERY**
UP 118.3%
0 NO CHANGE
UP 6.1%

**MOTOR VEHICLES and EQUIPMENT**
UP 195.7%
UP 50.0%
UP 8.9%

[1] Data: Federal Trade Commission–Securities and Exchange Commission.
[2] Data: U.S. Dept. of Commerce and Securities and Exchange Commission; seasonally adjusted.
[3] Average hourly earnings of production workers. Data: U.S. Dept. of Labor.

31

# VI. Goals And Programs For Wage Improvement

To summarize what has been said: More employment and less unemployment depend upon more demand for goods and services. The biggest factor in demand is consumer spending, and the biggest weakness in consumer spending today is the wage deficiency. It follows that efforts to lift wages should have top priority in any effective economic program.

## The size of the wage-lift task

The chart on page 35, based upon a complete model of conditions essential to full prosperity, depicts needed changes in the main sectors of the economy. Looking at the three large components in total national production—consumer spending, gross private investment, and public outlays for goods and services—it appears that in the neighborhood of two-thirds of the needed advances in total national production would have to take the form of advances in consumer spending. Looking next at the components in consumer income, it appears that the bulk of the needed advances in these incomes would have to take the form of advances in wages and salaries. Starting with 1953 as a base, wages and salaries need to rise by about 72 billion dollars by 1966, and by 160 billion by 1970. This requires a very much more rapid advance than during recent years.

Much of this wage advance can come from additional employment. But to bring this about, there must first be advances in the incomes and spending power of those now employed. This requires increases in hourly wage rates, large and prompt enough to close the growing gap between wage rate increases and productivity gains, and to reflect productivity-potential gains under the new technology and automation. The magnitudes are indicated by the charts on pages 27 and 28 of the preceding chapter.

Adequate wage rate increases will require vigorous collective bargaining; further organization of workers; recognition by management of the vital need to expand wage purchasing power; and Government reconsideration of its wage guidelines.

## Needed improvements in minimum wage legislation

The victims of excessively low wages are not rescued promptly by the general advance of wage rates. The Federal minimum wage floor of $1.25 an hour is much too low. This minimum wage floor should be lifted to $2.00 an hour, and coverage extended to many millions of workers now excluded. Even $2.00 an hour would provide, even with year-round employment, an income somewhat below the level needed to lift an average-sized American family out of poverty, and would be about $2,000 below the level needed to lift it above deprivation.

## The wage problem and the shorter work week

Very few deny that full employment with the standard work week of today would be better than full employment with a standard work week reduced to 35 hours or thereabouts. Our Nation has enormous unmet needs, both private and public, and great responsibilities overseas. And if good working conditions are maintained, the standard work week of today is compatible with the health and general well-being of workers.

But we have been making precious little progress toward full employment with the standard work week as it now is. And many people in high places insist that expansion of demand sufficient to restore full employment with this standard work week is confronted by insuperable "political" barriers. This being the case, the reduction of the work week to 35 hours or thereabouts should have very high priority attention.

Full employment promoted by a reduced work week (although many other measures are also required) would be more efficient than very high unemployment with the standard work week as it now is. It is more humane to have people employed 35 hours a week than to have them unemployed in very large numbers. It costs the Nation less to reduce the work week than to suffer the wastes and dollar costs of high unemployment. Plant capacities more fully used through a shorter work week might be more efficiently run than plants with large idle capacities, and thus yield higher gains in productivity.

## The shorter work week would not be inflationary

Opponents of the shorter work week say that the reduction in working hours would be accompanied by maintenance of current weekly take-home pay, and that this, by increasing hourly wage rates, would force up prices and thus hurt us both at home and overseas. This whole argument neglects the injurious lag of hourly wage rates behind productivity trends, depicted in the previous chapter. This lag can be remedied by maintaining the length of the work week and raising weekly take-home pay, or by reducing the length of the work week and maintaining weekly take-home pay.

But whichever method is used, prices are not "forced" up by wage rate increases that merely keep pace with productivity gains. To the extent that such wage rate increases have been used at times as an excuse for higher prices, this is hurtful to the whole economy because it causes consumption to lag behind profits and investment. The need is for wage rate increases geared to productivity gains *and* a stable price level. Those who

33

claim success for official wage-price guidelines, in that prices have been stabilized while needed wage expansion has been sacrificed, neglect the central cause of our economic trouble: Despite stable prices, there has been and still is a bad balance between profits and wages, and correspondingly between investment and consumption. This bad balance, in more extreme form, developed in the 1920's, when prices were remarkably stable.

## Massive poverty in America is largely a wage problem

Most of the poor are low-paid wage earners, unemployed wage earners, young people without jobs who ought to be wage earners, or retired wage earners. Massive improvement on the wage front is thus essential to reduction of massive poverty in America.

As shown by the chart on page 36, 10.4 million multiple person families in 1961 and about 4 million unattached individuals had annual incomes below the level required to lift them out of poverty. These two groups contain about 38 million Americans, or about one-fifth of a nation. To be sure, many pin-pointed programs are needed to help these people. But all experience indicates that they will be helped most by sustained full employment and production. During 1939-1953, when these conditions were reasonably sustained, vast gains were made in the reduction of poverty. During the more than a decade since 1953, with low economic growth and high unemployment, reduction of poverty in America has been brought almost to a standstill. And among all the programs needed to restore and maintain full employment and production, none is more important than a progressive wage policy.*

## Other measures needed to improve incomes

Programs in addition to wage efforts are needed to improve the lot of those two-fifths of all Americans who are poor or deprived. The chart on page 35 indicates the size of these tasks. Starting from the base year 1963, transfer payments, especially to enlarge old-age insurance benefits, need to be lifted by 15.6 billion dollars by 1966, and by 29 billion by 1970. Net farm income needs to be lifted 14.2 billion dollars by 1966, and 19.2 billion by 1970. Federal public outlays for goods and services (calendar years) to expand especially programs related to education, health, housing, and welfare, need to be lifted 13 billion dollars by 1966, and 27 billion by 1970. The need for expansion of these public programs is discussed further in Chapter IX.

The two following charts complete this chapter.

---

* For full discussion, see the Conference study, *Poverty and Deprivation in the U.S.*

# GOALS FOR 1966 AND 1970, PROJECTED
# FROM ACTUAL LEVELS IN 1963 [1]

Dollar Figures in 1962 Dollars

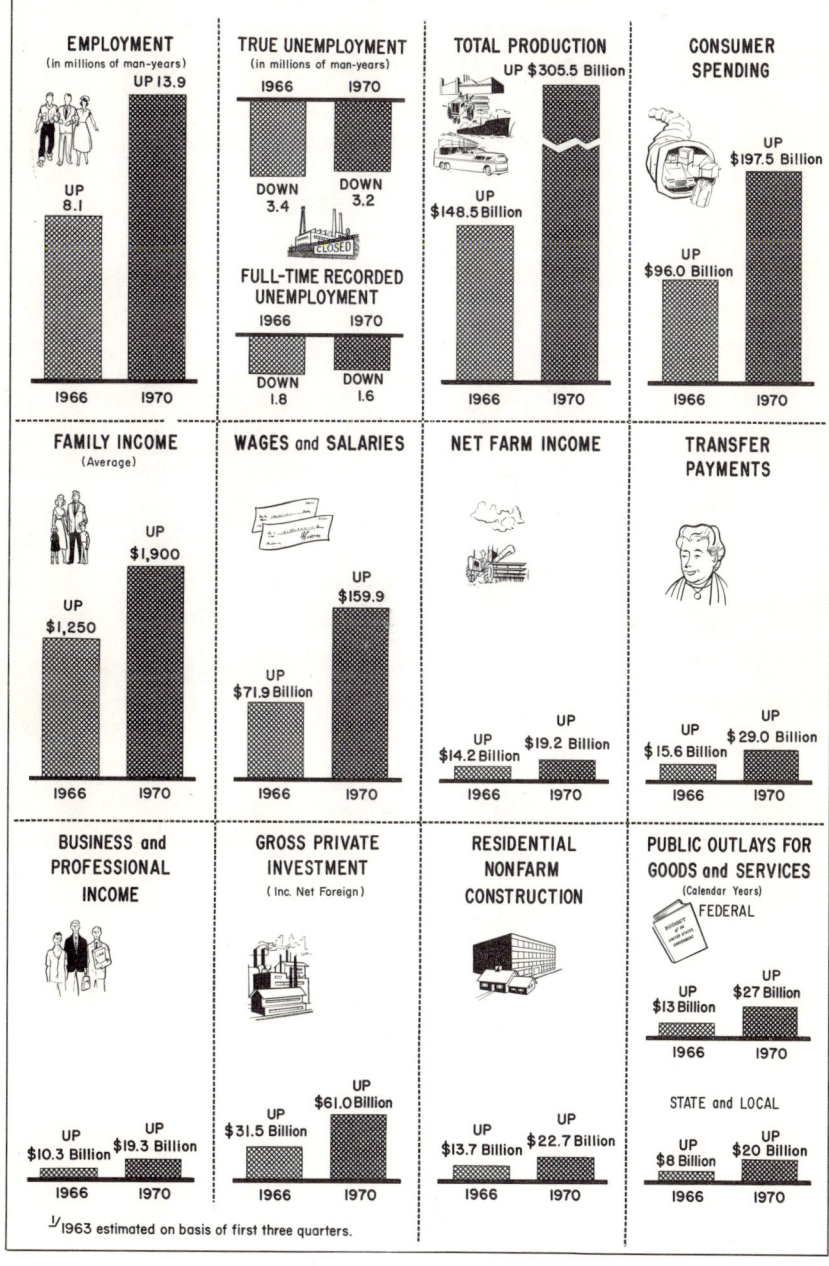

**EMPLOYMENT**
(in millions of man-years)
UP 13.9
UP 8.1
1966    1970

**TRUE UNEMPLOYMENT**
(in millions of man-years)
1966    1970
DOWN 3.4    DOWN 3.2

**FULL-TIME RECORDED UNEMPLOYMENT**
1966    1970
DOWN 1.8    DOWN 1.6

**TOTAL PRODUCTION**
UP $305.5 Billion
UP $148.5 Billion
1966    1970

**CONSUMER SPENDING**
UP $197.5 Billion
UP $96.0 Billion
1966    1970

**FAMILY INCOME**
(Average)
UP $1,900
UP $1,250
1966    1970

**WAGES and SALARIES**
UP $159.9
UP $71.9 Billion
1966    1970

**NET FARM INCOME**
UP $19.2 Billion
UP $14.2 Billion
1966    1970

**TRANSFER PAYMENTS**
UP $29.0 Billion
UP $15.6 Billion
1966    1970

**BUSINESS and PROFESSIONAL INCOME**
UP $19.3 Billion
UP $10.3 Billion
1966    1970

**GROSS PRIVATE INVESTMENT**
( Inc. Net Foreign )
UP $61.0 Billion
UP $31.5 Billion
1966    1970

**RESIDENTIAL NONFARM CONSTRUCTION**
UP $22.7 Billion
UP $13.7 Billion
1966    1970

**PUBLIC OUTLAYS FOR GOODS and SERVICES**
(Calendar Years)
FEDERAL
UP $27 Billion
UP $13 Billion
1966    1970

STATE and LOCAL
UP $20 Billion
UP $8 Billion
1966    1970

[1] 1963 estimated on basis of first three quarters.

35

# AMERICANS LIVING IN POVERTY AND THEIR SHARE OF INCOME, 1961

Annual Incomes, Before Taxes[1/]

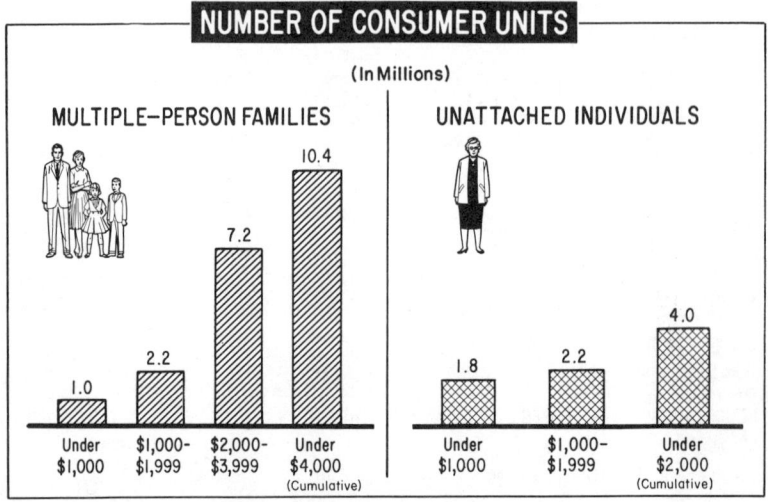

## NUMBER OF CONSUMER UNITS

(In Millions)

### MULTIPLE–PERSON FAMILIES

| Under $1,000 | $1,000–$1,999 | $2,000–$3,999 | Under $4,000 (Cumulative) |
|---|---|---|---|
| 1.0 | 2.2 | 7.2 | 10.4 |

### UNATTACHED INDIVIDUALS

| Under $1,000 | $1,000–$1,999 | Under $2,000 (Cumulative) |
|---|---|---|
| 1.8 | 2.2 | 4.0 |

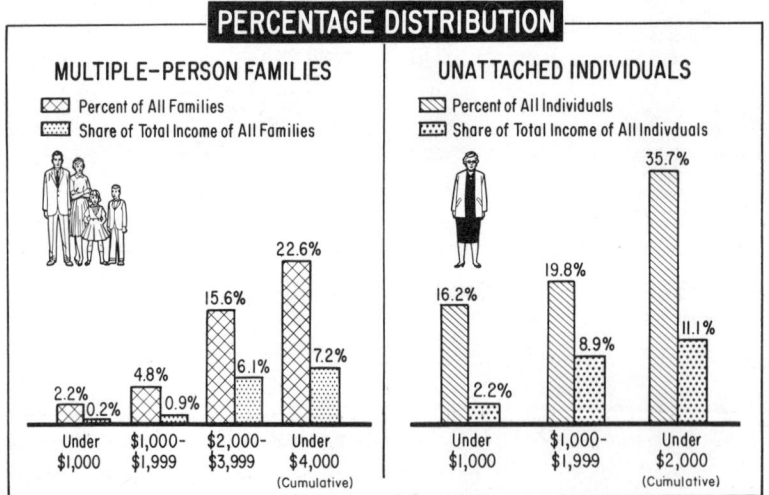

## PERCENTAGE DISTRIBUTION

### MULTIPLE–PERSON FAMILIES

Percent of All Families
Share of Total Income of All Families

| Under $1,000 | $1,000–$1,999 | $2,000–$3,999 | Under $4,000 (Cumulative) |
|---|---|---|---|
| 2.2% / 0.2% | 4.8% / 0.9% | 15.6% / 6.1% | 22.6% / 7.2% |

### UNATTACHED INDIVIDUALS

Percent of All Individuals
Share of Total Income of All Indivduals

| Under $1,000 | $1,000–$1,999 | Under $2,000 (Cumulative) |
|---|---|---|
| 16.2% / 2.2% | 19.8% / 8.9% | 35.7% / 11.1% |

[1/] Includes, in addition to cash income, the monetary value of food and fuel produced by farm families for their own use, and other nonmoney income.

Data: Department of Commerce, except that numbers of consumer units in "under $1,000" and "$1,000-$1,999" groupings are estimated by CEP on basis of Commerce Department data for families and individuals with incomes "under $2,000".

CEP has also estimated shares of income for "under $1,000" and "$1,000-$1,999" groupings.

# VII. The Changing Structure
# Of Employment Opportunity:
# The Problem Of The Structure of Demand

## Significance of the structure of demand

Generating enough total demand to restore and maintain full employment cannot be separated from the problem of the composition or structure of this total demand. For example, if industry A is registering much faster rates of change in technology and automation than industry B, similar rates of expansion of demand for the products of the two industries will result in less employment expansion in industry A than in industry B. Because any given amount of expansion of demand for products has different employment effects depending upon the points where this demand is exerted, the composition or structure of demand has great bearing upon the total volume of employment expansion.

The composition or structure of demand is important for other reasons. Theoretically, full employment could be achieved and maintained by building pyramids in the desert, or by increasing demand for nothing but consumer durables, or by a distribution of income which resulted in people at the top of the income structure building enough palaces and buying enough yachts to accomplish full employment, provided that they wanted to do so. But these kinds of full employment would give us useless products, or more of some products than we could possibly use, or outrage our sense of social justice.

This means that, in order to develop a composition and structure of demand which will yield sustained full employment, we must take account of relative trends in technology and automation in different sectors of the economy, the changing pattern of consumer tastes, and the relative size and priorities of currently unmet national needs.

## Impact of technology and automation

The chart on page 40 takes the period 1947-1949 as a base period when a given number of people employed for one year could turn out a given volume of physical production in that year. This base period 1947-1949 is thus taken to represent a ratio of employment to output represented by an index of 100. The chart shows that, by 1962, the ratio of employment to output in agriculture fell to 51.7. In other words, about half as many workers could turn out the same amount of products.

37

By 1962, the ratio of employment to output dropped to 51.4 in mining, and dropped to 62.5 in all manufacturing. The declines in various types of manufacturing were severe in each case, and particularly severe in the case of motor vehicles and other transportation equipment, where the index dropped to 47.5 in 1962. In railroads, the index dropped to 59.2 in 1962. In contract construction, the drop was relatively smaller but nonetheless substantial, to an index of 86.6.

## Employment trends, production workers in manufacturing

During 1947-1953, characterized generally by high overall economic growth, total employment of production workers in manufacturing, as shown by the chart on page 41, increased at an average annual rate of 1.3 percent, with increases at various rates in almost all of the main types of manufacturing. Thus, the high increases in demand for products more than compensated for the advances in technology. But during 1953-1963, characterized by a low rate of economic growth, total employment of production workers in manufacturing declined at an average annual rate of 1.1 percent. Thus, in terms of employment opportunity, the advances in technology and automation more than cancelled out the effects of the slow increase in demand for products. The same comment explains the average annual rate of employment decline of 2.5 percent in iron and steel, 2.7 percent in autos, trucks and parts, 3.9 percent in other transportation equipment, and 3.6 percent in petroleum refining and related products. In electrical machinery, there was a very slow rise in employment, indicating that the increasing demand for products more than compensated for the very high rate of technological change and automation.

## Employment trends, all nonagricultural wage and salary workers, and in the entire civilian economy

The chart on page 42 carries the examination of employment trends into the whole nonfarm economy, and deals with all wage and salary workers, not just production workers. Here again, the overall trends were very much more favorable during 1947-1953 than during 1953-1963, reflecting in the later period the poor performance of the whole economy and the relative rates of technological change and automation in the different sectors. During the later period, employment moved downward not only in manufacturing, but also in mining and in transportation and public utilities. The most favorable upward movements, in the order of relative rates of increase, were in government employment, the services and miscellaneous category, and finance, insurance, and real estate.

Trends in total civilian employment by occupation, 1947-1962, are shown by the chart on page 43. These trends appear to have been roughly consistent with the trends already discussed, allowing for some variations due to the differences among production-worker employment, wage- and salary-worker employment, and total employment.

## Unemployment rate trends, wage and salary workers

The chart on page 44 shows relative rates of unemployment in the different sectors. Of course, for the total civilian labor force and also in each sector, unemployment rates have averaged higher during 1953-1963 than during 1947-1953. The highest rate of unemployment during the later period, and also in 1963, has been in construction. This has great significance for the housing program subsequently to be discussed. The very high rates of unemployment in agriculture, and in forestry, fishing, and mining, identify the extraordinarily high rate of productivity advance in agriculture and mining relative to expansion of use of their products. The very high rate of unemployment in manufacturing also reflects the high progress of technology and automation, relative to the expansion of demand. The lowest rate of unemployment has been in public administration, reflecting the expanding role of the public services.

## The labor force adapts itself well to changes in the structure of demand

It is frequently said that the high unemployment of today is largely attributable to a substantial part of the labor force being unsuited for or untrained for the available types of jobs. Of course, there is always need for programs to educate, train and retrain, and relocate workers. But every period of reasonably full employment—and above all the World War II period—has taught us that fitting the worker to the job is a relatively minor problem when the jobs are there. Many workers are trained on the job, and we cannot know what kind of jobs to train people for until the jobs are available or being made available. Further, as shown earlier, the general and widespread nature of high unemployment makes it very clear that the main trouble is the shortage of total jobs, rather than the unadaptability of the labor force. Thus, if the composition and structure of demand receive due attention, the structure of the labor force will adjust itself without great difficulty.

The five following charts provide further details.

# RATIO OF VOLUME OF EMPLOYMENT TO PHYSICAL VOLUME OF PRODUCTION

(1947-1949 Ratio of Employment to Production = 100)

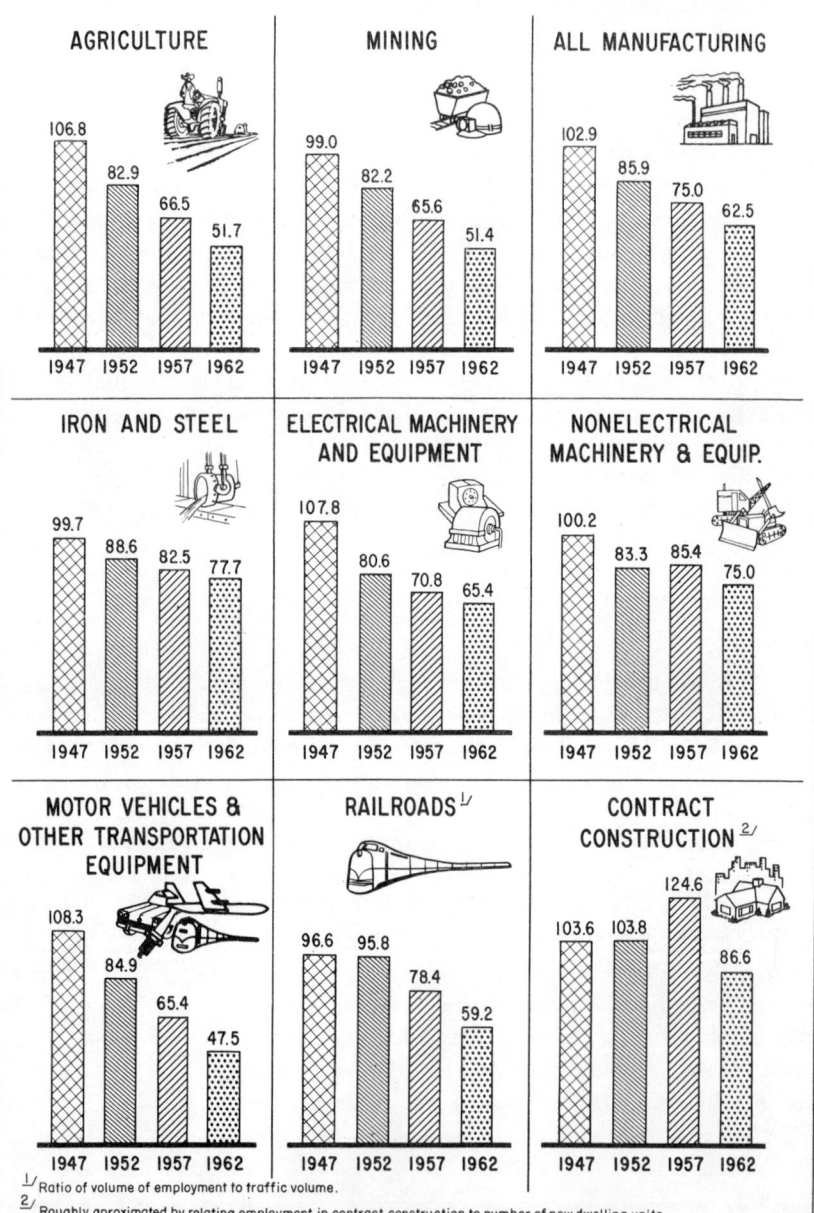

### AGRICULTURE

| 1947 | 1952 | 1957 | 1962 |
|------|------|------|------|
| 106.8 | 82.9 | 66.5 | 51.7 |

### MINING

| 1947 | 1952 | 1957 | 1962 |
|------|------|------|------|
| 99.0 | 82.2 | 65.6 | 51.4 |

### ALL MANUFACTURING

| 1947 | 1952 | 1957 | 1962 |
|------|------|------|------|
| 102.9 | 85.9 | 75.0 | 62.5 |

### IRON AND STEEL

| 1947 | 1952 | 1957 | 1962 |
|------|------|------|------|
| 99.7 | 88.6 | 82.5 | 77.7 |

### ELECTRICAL MACHINERY AND EQUIPMENT

| 1947 | 1952 | 1957 | 1962 |
|------|------|------|------|
| 107.8 | 80.6 | 70.8 | 65.4 |

### NONELECTRICAL MACHINERY & EQUIP.

| 1947 | 1952 | 1957 | 1962 |
|------|------|------|------|
| 100.2 | 83.3 | 85.4 | 75.0 |

### MOTOR VEHICLES & OTHER TRANSPORTATION EQUIPMENT

| 1947 | 1952 | 1957 | 1962 |
|------|------|------|------|
| 108.3 | 84.9 | 65.4 | 47.5 |

### RAILROADS [1]

| 1947 | 1952 | 1957 | 1962 |
|------|------|------|------|
| 96.6 | 95.8 | 78.4 | 59.2 |

### CONTRACT CONSTRUCTION [2]

| 1947 | 1952 | 1957 | 1962 |
|------|------|------|------|
| 103.6 | 103.8 | 124.6 | 86.6 |

[1] Ratio of volume of employment to traffic volume.

[2] Roughly aproximated by relating employment in contract construction to number of new dwelling units.

# EMPLOYMENT TRENDS:
## PRODUCTION WORKERS, 1947-1963 [1/]
### (Average Annual Rates of Change)

**ALL MANUFACTURING**

UP 1.3% — 1953-'63 [1/]
1947-'53 — DOWN 1.1%

**IRON AND STEEL**

UP 1.3% — 1953-'63 [1/]
1947-'53 — DOWN 2.5%

**NONFERROUS METALS**

UP 0.7% — 1953-'63 [1/]
1947-'53 — DOWN 1.0%

**NONELECTRICAL MACHINERY**

UP 1.4% — 1953-'63 [1/]
1947-'53 — DOWN 1.2%

**ELECTRICAL MACHINERY**

UP 4.1% — 1947-'53
UP 0.3% — 1953-'63 [1/]

**AUTOS, TRUCKS AND PARTS**

UP 2.8% — 1947-'53
1953-'63 [1/] — DOWN 2.7%

**OTHER TRANSPORT. EQUIP.**

UP 11.7% — 1947-'53
1953-'63 [1/] — DOWN 3.9%

**FABRICATED METAL PRODUCTS**

UP 2.1% — 1947-'53
1953-'63 [1/] — DOWN 0.6%

**CHEMICALS AND ALLIED PRODUCTS**

UP 1.2% — 1947-'53
no change — 1953-'63

**PAPER AND ALLIED PRODUCTS**

UP 1.5% — 1947-'53
UP 1.0% — 1953-'63 [1/]

**RUBBER AND MISC. PLASTIC PROD.**

UP 1.5% — 1947-'53
UP 0.9% — 1953-'63 [1/]

**STONE, CLAY AND GLASS**

UP 0.8% — 1947-'53
1953-'63 [1/] — DOWN 0.1%

**PETROLEUM AND RELATED PROD.**

UP 0.3% — 1947-'53
1953-'63 [1/] — DOWN 3.6%

**FOOD AND BEVERAGES**

1947-'53 — DOWN 0.8%
1953-'63 [1/] — DOWN 1.4%

**TEXTILES, APPAREL AND RELATED**

1947-'53 — DOWN 0.7%
1953-'63 [1/] — DOWN 1.1%

**ALL OTHER PRODUCTION WORKERS IN MANUFACTURING**

UP 0.6% — 1953-'63 [1/]
1947-'53 — DOWN 1.1%

[1/] 1963 estimated on basis of average of first ten months, seasonally adjusted.

41

# EMPLOYMENT TRENDS: NONAGRICULTURAL WAGE AND SALARY WORKERS, 1947-1963 [1/]

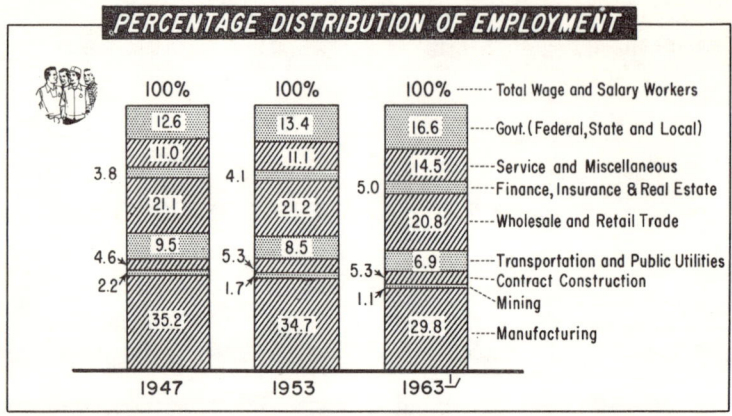

## PERCENTAGE DISTRIBUTION OF EMPLOYMENT

**1947** — 100%
- 12.6
- 11.0
- 3.8
- 21.1
- 4.6
- 9.5
- 2.2
- 35.2

**1953** — 100%
- 13.4
- 11.1
- 4.1
- 21.2
- 5.3
- 8.5
- 1.7
- 34.7

**1963** [1/] — 100%
- 16.6
- 14.5
- 5.0
- 20.8
- 5.3
- 6.9
- 1.1
- 29.8

Total Wage and Salary Workers
- Govt. (Federal, State and Local)
- Service and Miscellaneous
- Finance, Insurance & Real Estate
- Wholesale and Retail Trade
- Transportation and Public Utilities
- Contract Construction
- Mining
- Manufacturing

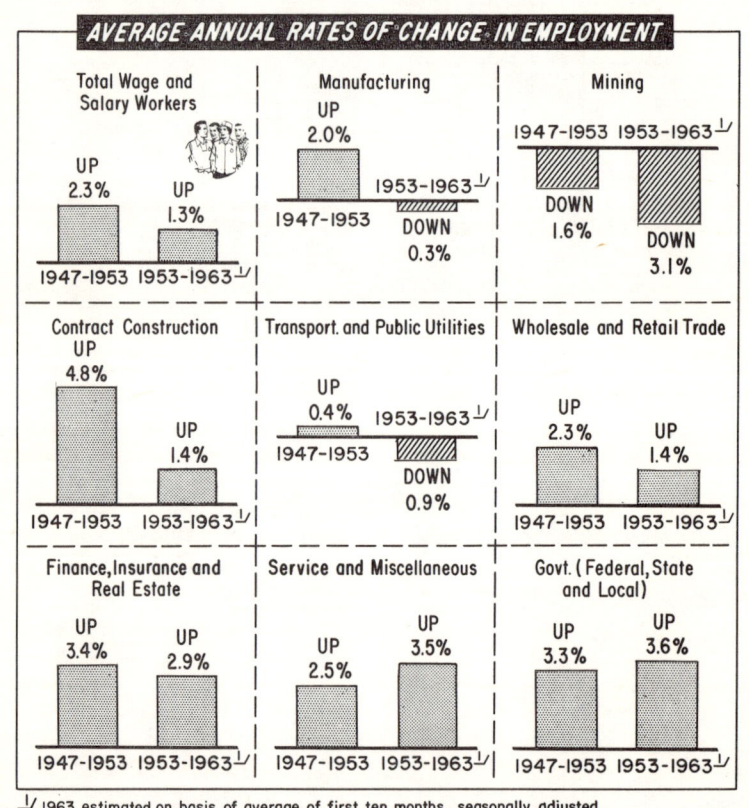

## AVERAGE ANNUAL RATES OF CHANGE IN EMPLOYMENT

**Total Wage and Salary Workers**
UP 2.3% (1947-1953) UP 1.3% (1953-1963 [1/])

**Manufacturing**
UP 2.0% (1947-1953) DOWN 0.3% (1953-1963 [1/])

**Mining**
(1947-1953) DOWN 1.6% (1953-1963 [1/]) DOWN 3.1%

**Contract Construction**
UP 4.8% (1947-1953) UP 1.4% (1953-1963 [1/])

**Transport. and Public Utilities**
UP 0.4% (1947-1953) DOWN 0.9% (1953-1963 [1/])

**Wholesale and Retail Trade**
UP 2.3% (1947-1953) UP 1.4% (1953-1963 [1/])

**Finance, Insurance and Real Estate**
UP 3.4% (1947-1953) UP 2.9% (1953-1963 [1/])

**Service and Miscellaneous**
UP 2.5% (1947-1953) UP 3.5% (1953-1963 [1/])

**Govt. (Federal, State and Local)**
UP 3.3% (1947-1953) UP 3.6% (1953-1963 [1/])

[1/] 1963 estimated on basis of average of first ten months, seasonally adjusted.

42

# TOTAL CIVILIAN EMPLOYMENT TRENDS, BY OCCUPATION, 1947-1962

## Average Annual Rates of Change

### AGRICULTURE

| 1947-1953 | 1953-1962 |
|-----------|-----------|
| 3.6% DOWN | 2.7% DOWN |

### PRIVATE HOUSEHOLDS

| UP 2.8% | UP 3.2% |
|---------|---------|
| 1947-1953 | 1953-1962 |

### MANUFACTURING

| UP 1.2% | 0.1% DOWN |
|---------|-----------|
| 1947-1953 | 1953-1962 |

### MINING

| UP 2.3% | 5.2% DOWN |
|---------|-----------|
| 1947-1953 | 1953-1962 |

### CONTRACT CONSTRUCTION

| UP 4.0% | UP 0.4% |
|---------|---------|
| 1947-1953 | 1953-1962 |

### TRANSPORTATION, COMMUNICATIONS AND PUBLIC UTILITIES

| UP 0.1% | 0.9% DOWN |
|---------|-----------|
| 1947-1953 | 1953-1962 |

### WHOLESALE AND RETAIL TRADE

| UP 1.2% | UP 1.3% |
|---------|---------|
| 1947-1953 | 1953-1962 |

### FINANCE, INSURANCE AND REAL ESTATE

| UP 2.2% | UP 3.6% |
|---------|---------|
| 1947-1953 | 1953-1962 |

### SERVICES

| UP 3.2% | UP 4.0% |
|---------|---------|
| 1947-1953 | 1953-1962 |

### GOVERNMENT

| UP 3.5% | UP 3.7% |
|---------|---------|
| 1947-1953 | 1953-1962 |

**TOTAL CIVILIAN EMPLOYMENT:** 1947-1953: UP 1.2%  1953-1962: UP 1.0%

# UNEMPLOYMENT RATE TRENDS
# WAGE AND SALARY WORKERS, 1947-1963 [1]

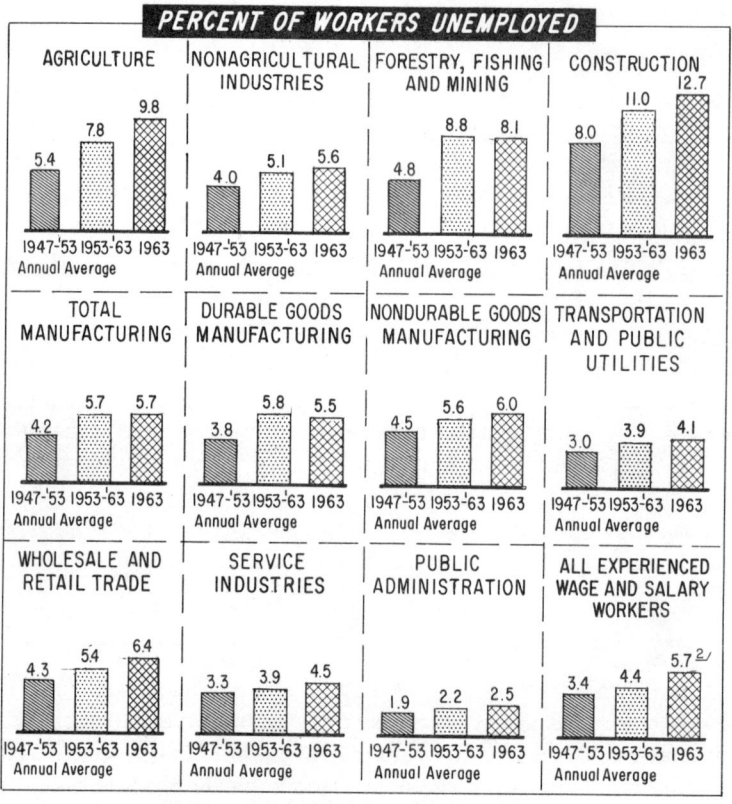

## PERCENT OF WORKERS UNEMPLOYED

AGRICULTURE
5.4 / 7.8 / 9.8
1947-'53 1953-'63 1963
Annual Average

NONAGRICULTURAL INDUSTRIES
4.0 / 5.1 / 5.6
1947-'53 1953-'63 1963
Annual Average

FORESTRY, FISHING AND MINING
4.8 / 8.8 / 8.1
1947-'53 1953-'63 1963
Annual Average

CONSTRUCTION
8.0 / 11.0 / 12.7
1947-'53 1953-'63 1963
Annual Average

TOTAL MANUFACTURING
4.2 / 5.7 / 5.7
1947-'53 1953-'63 1963
Annual Average

DURABLE GOODS MANUFACTURING
3.8 / 5.8 / 5.5
1947-'53 1953-'63 1963
Annual Average

NONDURABLE GOODS MANUFACTURING
4.5 / 5.6 / 6.0
1947-'53 1953-'63 1963
Annual Average

TRANSPORTATION AND PUBLIC UTILITIES
3.0 / 3.9 / 4.1
1947-'53 1953-'63 1963
Annual Average

WHOLESALE AND RETAIL TRADE
4.3 / 5.4 / 6.4
1947-'53 1953-'63 1963
Annual Average

SERVICE INDUSTRIES
3.3 / 3.9 / 4.5
1947-'53 1953-'63 1963
Annual Average

PUBLIC ADMINISTRATION
1.9 / 2.2 / 2.5
1947-'53 1953-'63 1963
Annual Average

ALL EXPERIENCED WAGE AND SALARY WORKERS
3.4 / 4.4 / 5.7 [2]
1947-'53 1953-'63 1963
Annual Average

## HIGHEST UNEMPLOYMENT RATES, 1962

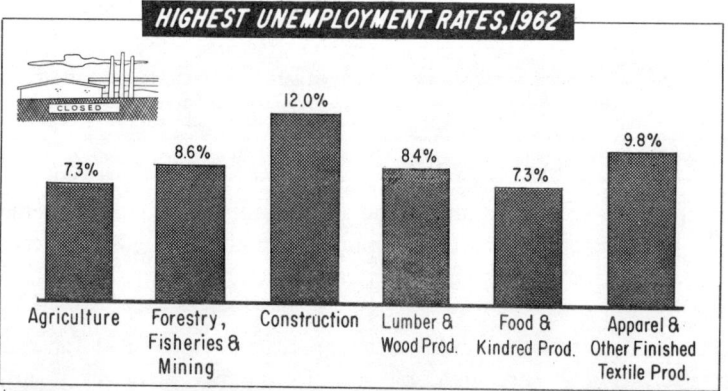

CLOSED

Agriculture 7.3% — Forestry, Fisheries & Mining 8.6% — Construction 12.0% — Lumber & Wood Prod. 8.4% — Food & Kindred Prod. 7.3% — Apparel & Other Finished Textile Prod. 9.8%

[1] 1963 estimated on basis of first nine months, not seasonally adjusted. Full-time unemployment only. 1957-1963 data are based on revised definition of unemployment and hence are not exactly comparable with earlier data.

[2] Seasonally adjusted rate: 5.5%

44

# VIII. Employment Goals,
# Looking Toward 1966 And 1970

### Difference between forecasts and goals

The employment goals projected in this chapter for 1966 and 1970 are not *forecasts* of what might happen automatically, or through the operation of "natural" economic forces. The whole purpose of this study is to show that, if we merely rely upon forecasts or these "natural" forces, we shall achieve neither full employment nor a desirable structure of employment and production.

Instead, the projections are practical *goals* for achieving, through proper changes in policies, maximum employment by early 1965, and maintaining it in the years thereafter. These goals also embody a composition or structure of employment which takes account of relative trends in technology and automation; the changing patterns of consumer needs and tastes; a proper balance between investment in the expansion of producer facilities and expansion of private and public consumption; and the competing priorities of our great national needs.

### Projected increases in civilian employment, by sectors

Taking account of all these factors, the chart on page 47, using 1962 as a base year, projects an increase in total civilian employment of 13.4 percent by 1966, and 22 percent by 1970.

With respect to the composition or structure of employment, a continuing downward trend in agricultural employment is projected. This is based upon the estimated consequences of fantastic changes taking place in agricultural technology, even after allowing for greatly expanded farm product exports, and for the much higher domestic consumption of farm products which would result from population growth and from improvements in the living standards of the two-fifths of all Americans now living in poverty or deprivation.

For manufacturing, there is projected a considerable advance in employment, reversing the downtrend of recent years. This is predicated upon enough expansion of consumption of manufactured products in a fully growing economy to more than compensate in terms of employment for the continuing advance of technology and automation.

The highest rate of employment expansion is projected for contract construction, and a very high rate also for finance, insurance, and real estate. This takes account of the relatively lower rates of technological change in these areas, and the immense housing and urban renewal needs subsequently discussed.

Very high rates of employment advance are also projected for the services and miscellaneous category, and for government employment. These allow for the relative shift toward services in a more productive economy with higher living standards, and the immense need for expanded public services discussed in the next chapter.

In accord with these relative rates of advance in employment in the various sectors, the distribution of total civilian employment would change considerably. For example, as shown by the chart on page 48, the downward trend in the percentage of total civilian employment in manufacturing would continue, and the same would be true of agriculture. The percentages engaged in the services and miscellaneous category, in government work, and in contract construction, would rise.

This portrayal of employment goals is embodied in the two following charts.

# GOALS FOR TOTAL CIVILIAN EMPLOYMENT, BY OCCUPATION, 1962-1966 AND 1962-1970

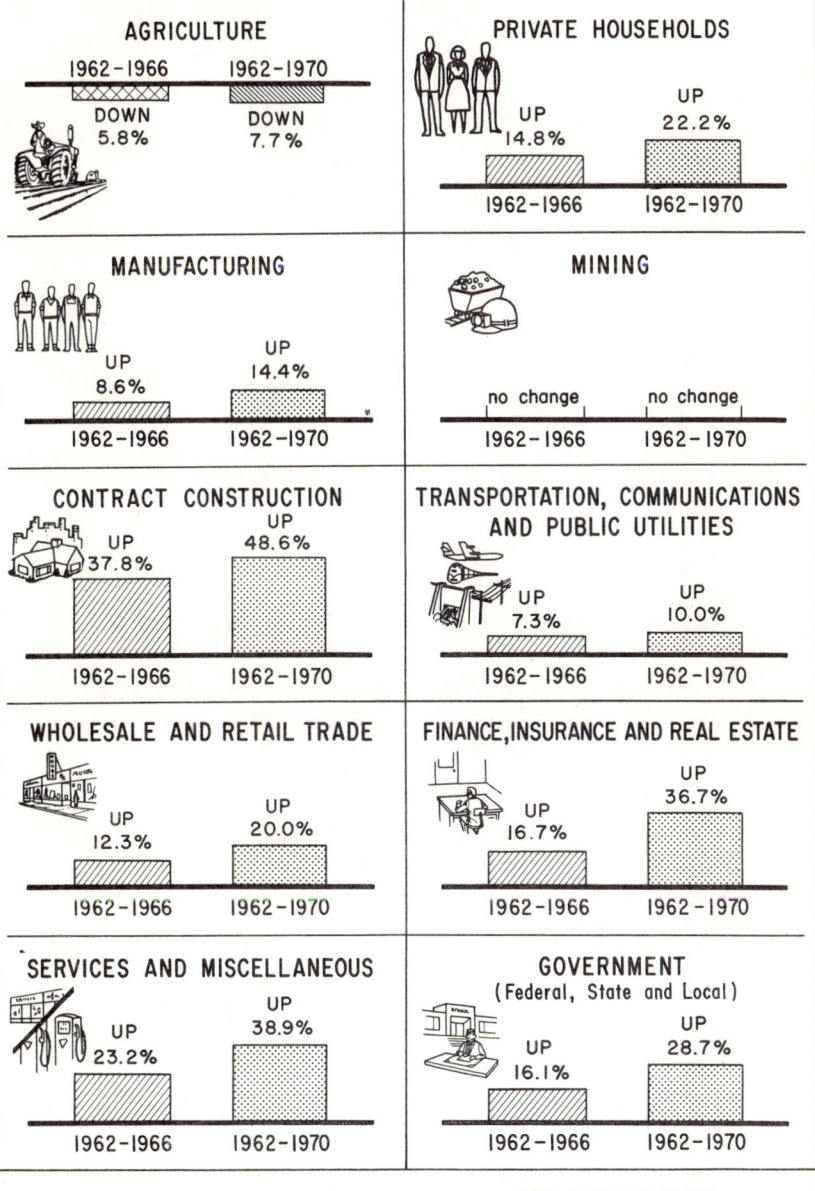

### AGRICULTURE

1962-1966 — DOWN 5.8%
1962-1970 — DOWN 7.7%

### PRIVATE HOUSEHOLDS

UP 14.8% — 1962-1966
UP 22.2% — 1962-1970

### MANUFACTURING

UP 8.6% — 1962-1966
UP 14.4% — 1962-1970

### MINING

no change — 1962-1966
no change — 1962-1970

### CONTRACT CONSTRUCTION

UP 37.8% — 1962-1966
UP 48.6% — 1962-1970

### TRANSPORTATION, COMMUNICATIONS AND PUBLIC UTILITIES

UP 7.3% — 1962-1966
UP 10.0% — 1962-1970

### WHOLESALE AND RETAIL TRADE

UP 12.3% — 1962-1966
UP 20.0% — 1962-1970

### FINANCE, INSURANCE AND REAL ESTATE

UP 16.7% — 1962-1966
UP 36.7% — 1962-1970

### SERVICES AND MISCELLANEOUS

UP 23.2% — 1962-1966
UP 38.9% — 1962-1970

### GOVERNMENT
(Federal, State and Local)

UP 16.1% — 1962-1966
UP 28.7% — 1962-1970

**TOTAL CIVILIAN EMPLOYMENT:** 1962-1966: UP 13.4%
1962-1970: UP 22.0%

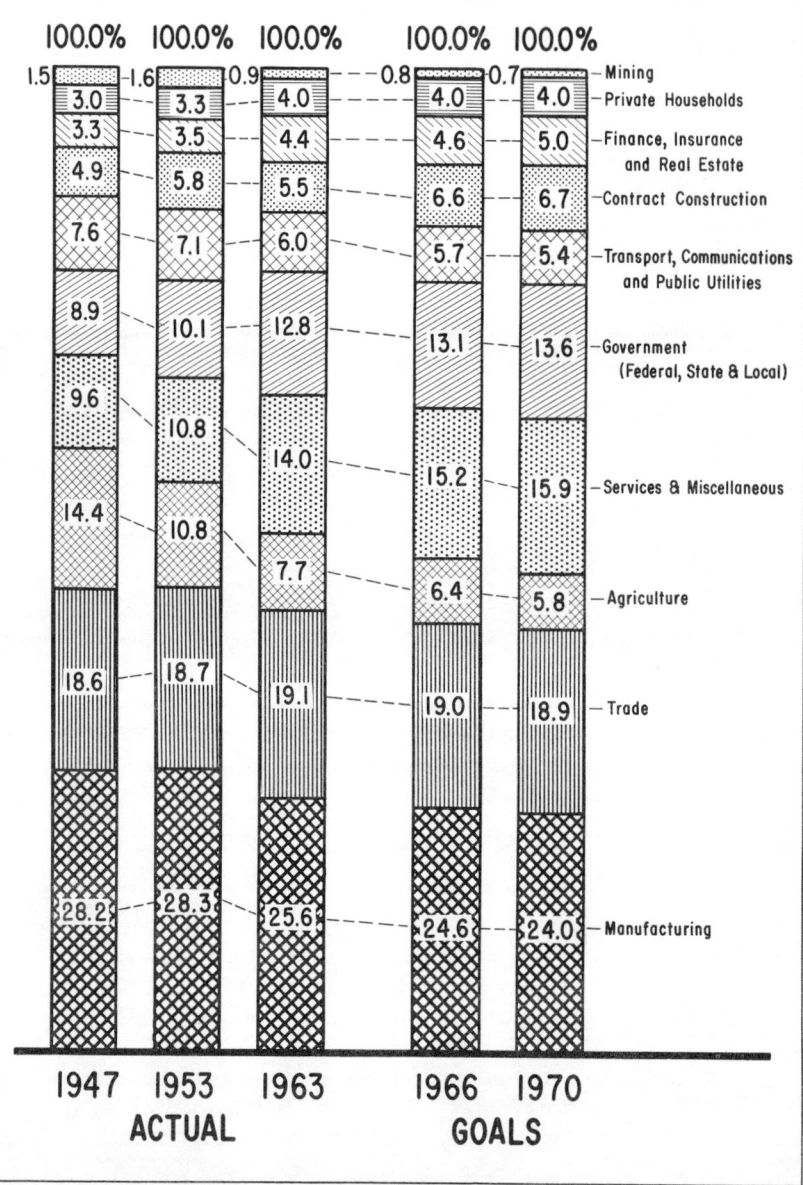

# TOTAL CIVILIAN EMPLOYMENT
## DISTRIBUTION 1947, 1953, AND 1962,
## AND GOALS FOR 1966 AND 1970

# IX. The Imperative Need
# For More Federal Public Spending

## General summation of the need

Restoration and maintenance of maximum employment and production require, as we have noted, much larger increases in Federal outlays than are now under consideration. This crucial point requires summation.

*First,* as shown by the chart on page 17 of an earlier chapter, a 4.2 billion dollar average annual deficiency in Federal goods and services outlays, especially taking account of the "multiplier" effects of Federal spending, has been a very important factor in the chronic rise of unemployment and idle plant during the period 1953-1963.

*Second,* as shown by the chart on page 35 of an earlier chapter, a balanced model for restoration and maintenance of maximum employment and production requires that Federal public outlays for goods and services be lifted above the calendar 1963 level by 13 billion dollars in calendar 1966, and 27 billion in calendar 1970.

*Third,* as shown by the chart on page 47 of the preceding chapter, very large increases in public employment are needed to make the structure of employment responsive to technological trends, to the changing pattern of the people's wants, and to the priorities of our national needs.

*Fourth,* the required rapid expansion of some of the types of private employment shown by the same chart hinges upon large increases in public outlays. The huge projected expansion of employment in contract construction, especially housing and urban renewal, depends upon a vigorous admixture of additional private and public spending. The large projected expansion of private employment in the services category, especially in education, health, and recreation, calls for much more Federal assistance for building facilities and recruitment and training of personnel.*

## The specifics of the need

The chart on page 52 depicts goals for a Federal Budget geared to economic growth and public needs. The specific items include education, where per capita outlays (measured in 1962 dollars) should be lifted from $7.87 in fiscal 1964 to $32.86 by calendar 1970; health services and research, where Federal per capita outlays over the same period of time

---

* For more detailed discussion of Federal programs and their significance, see the Conference study *The Federal Budget and the "General Welfare."*

should be lifted from $8.41 to $22.54; public assistance, where the lift should be from $15.48 to $21.13; labor and manpower, and other welfare services, where the lift should be from $4.85 to $5.63; and housing and community development, where the lift should be from $1.42 to $15.49. For all domestic programs and services, Federal per capita outlays should be lifted from $186.83 to $239.44. All this allows for a lift in per capita Federal outlays for national defense, space technology, and all international, from $319.03 to $396.71.

Including other needs such as national resource development and conventional types of public works, the chart on page 53 shows that the total Federal Budget should be lifted from 98.8 billion dollars as originally proposed by the Administration for fiscal 1964 to 115 billion by calendar 1966, and 135.5 billion dollars by calendar 1970. In an economy growing at the rate which these outlays would help to induce, total Federal outlays as a percent of total national production would decline considerably, and the total national debt as a percent of total national production would decline greatly.

The time to start moving on these increased Federal outlays is now. The intolerable levels of unemployment and idle plant demands it; the needs of the people require it; and ordinary economic prudence calls for it. As a first step, it is recommended that outlays for domestic purposes in the Federal Budget for fiscal 1965, to be presented in January 1964, be 3.5 to 4 billion dollars higher than those originally presented for fiscal 1964. Allowing for built-in increases, and desirable expansion in some other areas including assistance overseas and space research and technology, the total Federal Budget for fiscal 1965 ought to be 8 to 8.5 billion higher than the original Budget for 1964, or about 107 billion.

## Tax reduction is no substitute for increased Federal spending

Tax reduction, by placing more spendable income in the hands of consumers, will increase demand for many types of production. But as earlier shown, any feasible increases in this kind of demand, combined with the rates of advance in technology and automation in the industries where these products are being turned out, do not offer reasonable promise of contributing a major portion of the estimated 22.5 million additional jobs required over the next decade (about one-half net additions, and about one-half needed in place of jobs eliminated by technology and automation). Nor can tax reduction alone make a major contribution to many of the highest priorities of our national public needs.

50

## Specific defects in the pending tax bill

The distribution of tax cuts under the tax bill pending as of this writing, as shown by the chart on page 54, is weighted far too heavily on the side of investment stimulation, and sorely neglectful of the dominant need to stimulate consumption. And as shown by the chart on page 55, the distribution of the personal tax cuts is slanted far too heavily in favor of high income groups, with corresponding neglect of those lower down in the income structure. This is economically undesirable for reasons developed throughout this study; it is also extremely inequitable.

And finally, the tax bill in its pending form would provide but small stimulus to the economy, relative to the need. This study estimates that the tax bill as now written would have a stimulative effect of only about 30 billion dollars upon total national production when it became fully effective as of 1965 or 1966. This would be much less than one-third of the difference between needed total national production in 1966 and the likely level if the tax cuts were not put into effect. For the first calendar year in which the tax proposal would be applied, presumably 1964, this study estimates that its stimulative effect upon total national production would be in the neighborhood of 14.5 billion dollars, or only about one-fourth of the difference between needed total national production in 1964 and the likely level if the tax proposal were not enacted.

## Proposed revisions in the tax bill

For these reasons, it is highly desirable that the pending tax bill be revised. The corporate tax cuts are unnecessary and would be substantially wasted, because investors both corporate and personal are now amply supplied with funds, and because any inadequacy in investment is due only to deficient private consumption and public outlays. In addition, the pending tax bill allocates about twice as much tax reduction as would seem desirable to the top income eighth of all taxpayers. This excessive allocation, plus the corporate tax cuts, come to the neighborhood of 4 billion dollars. This portion of the tax cut ought to be used instead to lift the standard exemptions. By thus redirecting much more of the tax cut to those lower down in the income structure who would spend it promptly for consumption, the stimulative effect upon the economy would be much greater. This change would accord with social justice.*

The four following charts complete this chapter.

---

* For comprehensive treatment of the whole problem of taxation and of the pending tax bill, see the Conference study *"Taxes and the Public Interest."*

51

# GOALS FOR A FEDERAL BUDGET GEARED TO ECONOMIC GROWTH AND PUBLIC NEEDS

### 1964, Fiscal Year; 1966 and 1970, Calendar Years
#### Per Capita Outlay in 1962 Dollars

## TOTAL FEDERAL OUTLAYS

| Year | % of Total Output | $ Per Capita |
|---|---|---|
| 1964 Adm.[1/] | 16.80 | $505.86 |
| 1966 Goal | 15.91 | 577.89 |
| 1970 Goal | 15.41 | 636.15 |

## NATIONAL DEFENSE, SPACE TECHNOLOGY AND ALL INTERNATIONAL

| Year | % of Total Output | $ Per Capita |
|---|---|---|
| 1964 Adm.[1/] | 10.59 | $319.03 |
| 1966 Goal | 9.82 | 356.78 |
| 1970 Goal | 9.61 | 396.71 |

## EDUCATION

| Year | % of Total Output | $ Per Capita |
|---|---|---|
| 1964 Adm.[1/] | .26 | $7.87 |
| 1966 Goal | .62 | 22.61 |
| 1970 Goal | .80 | 32.86 |

## HEALTH SERVICES AND RESEARCH

| Year | % of Total Output | $ Per Capita |
|---|---|---|
| 1964 Adm.[1/] | .28 | $8.41 |
| 1966 Goal | .44 | 16.08 |
| 1970 Goal | .55 | 22.54 |

## PUBLIC ASSISTANCE

| Year | % of Total Output | $ Per Capita |
|---|---|---|
| 1964 Adm.[1/] | .51 | $15.48 |
| 1966 Goal | .51 | 18.59 |
| 1970 Goal | .51 | 21.13 |

## LABOR AND MANPOWER, AND OTHER WELFARE SERVICES

| Year | % of Total Output | $ Per Capita |
|---|---|---|
| 1964 Adm.[1/] | .16 | $4.85 |
| 1966 Goal | .15 | 5.53 |
| 1970 Goal | .14 | 5.63 |

## HOUSING AND COMMUNITY DEVELOPMENT

| Year | % of Total Output | $ Per Capita |
|---|---|---|
| 1964 Adm.[1/] | .05 | $1.42 |
| 1966 Goal | .31 | 11.06 |
| 1970 Goal | .37 | 15.49 |

## ALL DOMESTIC PROGRAMS AND SERVICES

(Includes also Agriculture; Natural Resources; Veterans; Commerce; Interest; General Government, etc.)

| Year | % of Total Output | $ Per Capita |
|---|---|---|
| 1964 Adm.[1/] | 6.21 | $186.83 |
| 1966 Goal | 6.09 | 221.11 |
| 1970 Goal | 5.80 | 239.44 |

[1/] Administration's proposed budget as of Jan. 17, 1963.

# TOWARD A FEDERAL BUDGET CONSISTENT WITH MAXIMUM EMPLOYMENT AND THE PRIORITIES OF NATIONAL PUBLIC NEEDS

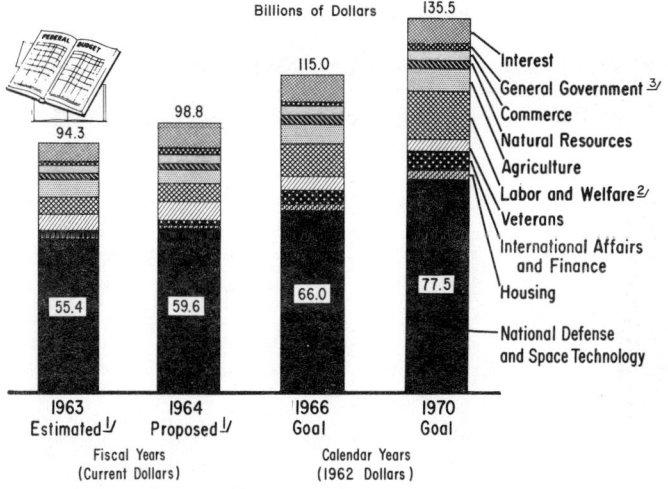

Billions of Dollars

Interest
General Government 3/
Commerce
Natural Resources
Agriculture
Labor and Welfare 2/
Veterans
International Affairs and Finance
Housing
National Defense and Space Technology

| 1963 Estimated 1/ | 1964 Proposed 1/ | 1966 Goal | 1970 Goal |
|---|---|---|---|
| Fiscal Years (Current Dollars) | | Calendar Years (1962 Dollars) | |

# BURDEN OF FEDERAL OUTLAYS IN A FULLY GROWING ECONOMY WOULD BE LOWER THAN IN RECENT YEARS

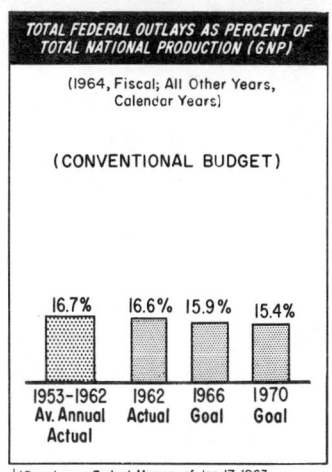

**TOTAL FEDERAL OUTLAYS AS PERCENT OF TOTAL NATIONAL PRODUCTION (GNP)**

(1964, Fiscal; All Other Years, Calendar Years)

(CONVENTIONAL BUDGET)

16.7%   16.6%   15.9%   15.4%

| 1953-1962 Av. Annual Actual | 1962 Actual | 1966 Goal | 1970 Goal |
|---|---|---|---|

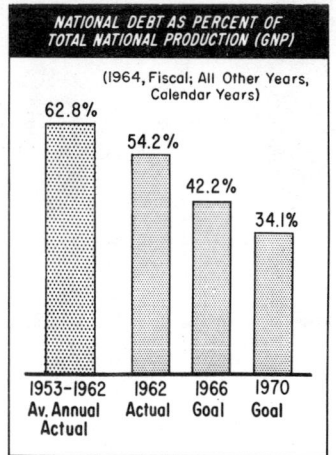

**NATIONAL DEBT AS PERCENT OF TOTAL NATIONAL PRODUCTION (GNP)**

(1964, Fiscal; All Other Years, Calendar Years)

62.8%   54.2%   42.2%   34.1%

| 1953-1962 Av. Annual Actual | 1962 Actual | 1966 Goal | 1970 Goal |
|---|---|---|---|

1/ Based upon Budget Message of Jan. 17, 1963

2/ Including education and health services

3/ Including contingencies and less interfund transactions

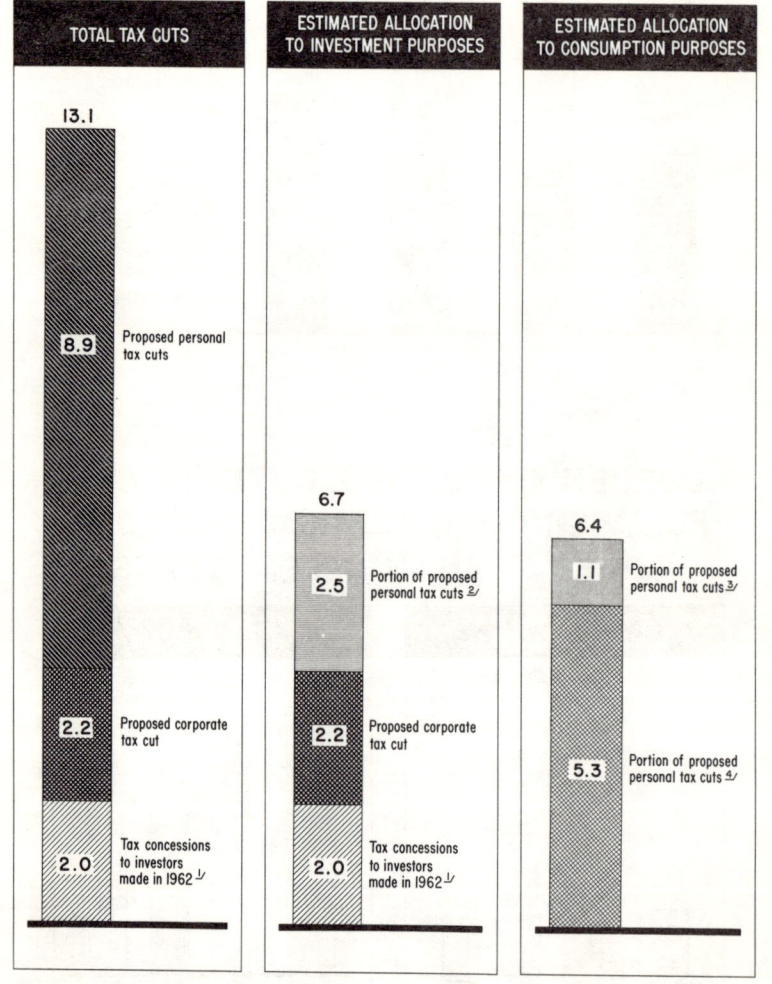

# PENDING TAX BILL: ESTIMATED DIVISION BETWEEN CUTS FOR INVESTMENT PURPOSES AND CUTS FOR CONSUMPTION PURPOSES

(Including-Tax Cuts of 1962)

Billions of Dollars

| TOTAL TAX CUTS | ESTIMATED ALLOCATION TO INVESTMENT PURPOSES | ESTIMATED ALLOCATION TO CONSUMPTION PURPOSES |
|---|---|---|

**TOTAL TAX CUTS**

13.1

8.9 — Proposed personal tax cuts

2.2 — Proposed corporate tax cut

2.0 — Tax concessions to investors made in 1962 [1]

**ESTIMATED ALLOCATION TO INVESTMENT PURPOSES**

6.7

2.5 — Portion of proposed personal tax cuts [2]

2.2 — Proposed corporate tax cut

2.0 — Tax concessions to investors made in 1962 [1]

**ESTIMATED ALLOCATION TO CONSUMPTION PURPOSES**

6.4

1.1 — Portion of proposed personal tax cuts [3]

5.3 — Portion of proposed personal tax cuts [4]

[1] Through Congressional and Executive action.

[2] Estimated portion of personal tax cuts, for those with incomes of $10,000 and over, which they would save for investment purposes.

[3] Estimated portion of personal tax cuts, for those with incomes of $10,000 and over, which they would spend for consumption.

[4] Personal tax cuts for those with incomes under $10,000.

Note: Estimates of division, CEP.

54

# PENDING TAX BILL, PERSONAL TAX CUTS
### Percent Tax Cut And Percent Gain In After-Tax Income
### Married Couple With Two Children At Various Income Levels [1]

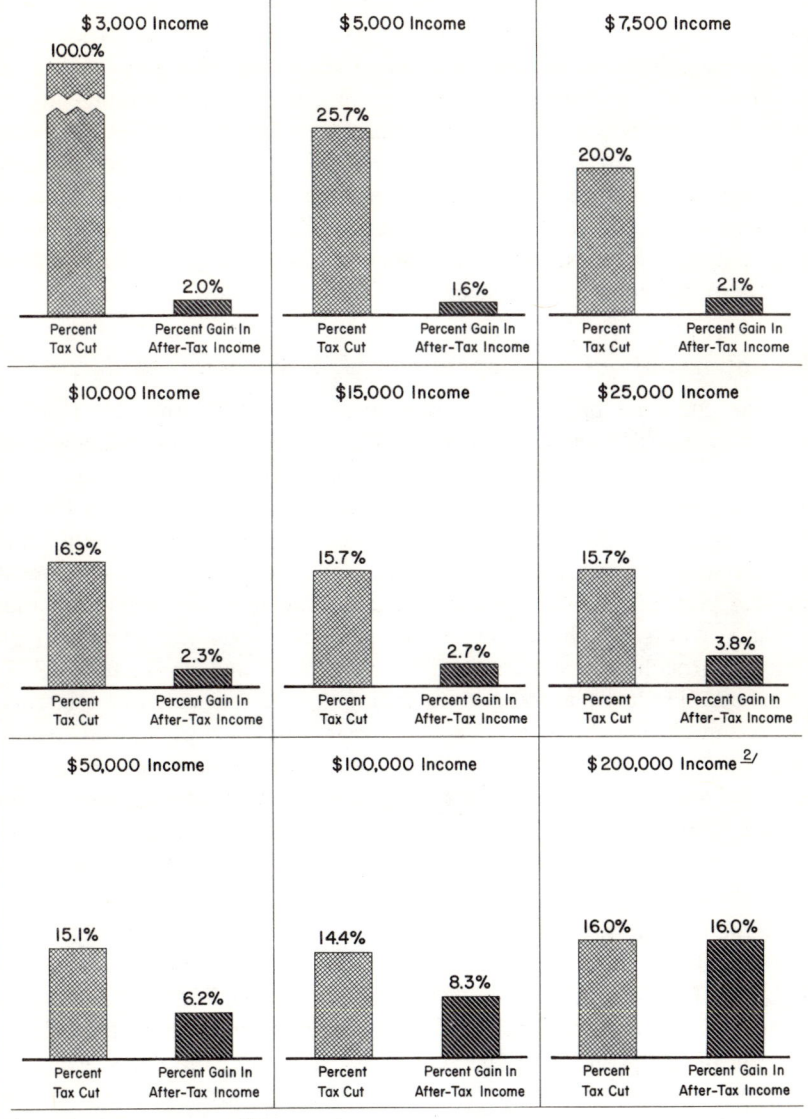

**$3,000 Income**

100.0%

2.0%

Percent Tax Cut · Percent Gain In After-Tax Income

**$5,000 Income**

25.7%

1.6%

Percent Tax Cut · Percent Gain In After-Tax Income

**$7,500 Income**

20.0%

2.1%

Percent Tax Cut · Percent Gain In After-Tax Income

**$10,000 Income**

16.9%

2.3%

Percent Tax Cut · Percent Gain In After-Tax Income

**$15,000 Income**

15.7%

2.7%

Percent Tax Cut · Percent Gain In After-Tax Income

**$25,000 Income**

15.7%

3.8%

Percent Tax Cut · Percent Gain In After-Tax Income

**$50,000 Income**

15.1%

6.2%

Percent Tax Cut · Percent Gain In After-Tax Income

**$100,000 Income**

14.4%

8.3%

Percent Tax Cut · Percent Gain In After-Tax Income

**$200,000 Income [2]**

16.0%

16.0%

Percent Tax Cut · Percent Gain In After-Tax Income

[1] Adjusted gross income levels.　[2] Estimated

Note: Standard deductions for $3,000 income level. Typical itemized deductions for other income levels.

# X. Economic Need
# For Greatly Expanded Housing Efforts

The discussion up to this point converges upon these conclusions: (a) a tremendous expansion of total demand for goods and services is needed to restore and maintain maximum employment and production; (b) the onrushing technology and automation, the changing pattern of consumer wants and needs, and the top priorities of our national needs, require substantial changes in the composition and structure of this total demand; and (c) all of these considerations call for large increases not only in private investment and consumer spending, but also in public outlays, especially at the Federal level. The balance of this study deals with housing, as perhaps the most important single area of opportunity for attention to all these problems and satisfaction of all these needs.

## The importance of housing in the U. S. economy to date

The chart on page 58 shows no year from 1947 through 1962 in which total housing outlays fell below 12.7 percent of total national production (true also of years omitted on the chart). In the years from 1953 forward, housing outlays, except in three years, have run at 15 percent or higher of total national production. Over the same span of time, the dollar value of residential construction has ranged between 36.5 percent and 48.4 percent of total new construction. And it is very significant that the ratio averaged annually about 43 percent during the years of high economic growth and reasonably full employment and production, 1947-1953, but fell to less than 40 percent during the years of very low economic growth and chronically rising idleness of manpower and plant, 1953-1962. The ratio was even lower during 1957-1962. And these *ratios* do not tell the whole story, because the other outlays with which housing outlays are compared have also been too low. It should also be noted that total housing outlays by consumers have ranged between 13 and 16 percent of all consumer spending.

In addition to the inadequate volume, the exceptional instability of the housing industry has been, by common consent, a prime factor in the instability of the whole economy. During the period from 1953 forward, the index of our total national production never dropped more than about 2 percent in real terms from one year to the next, and the index of industrial production never dropped more than 7 percent from one year to the next. But the index of total housing starts dropped almost 16 percent from 1955 to 1956 and another 7 percent from 1956 to 1957, and dropped more than 16 percent from 1959 to 1960. The relationship between these tremendous declines in housing starts and the recessions of 1957-1958 and 1960-1961 are abundantly apparent.

## The role of housing in economic restoration

This study estimates that a deficiency of about 35-40 billion dollars in private investment in residential construction during the period 1953-1963 accounted for about a third of the deficiency in gross private investment, estimated in excess of 118 billion. As shown by the earlier chart on page 35, as part of a balanced program for economic restoration starting from a 1963 base, investment in new residential non-farm construction needs to be lifted more than 56 percent by 1966, and more than 93 percent by 1970. This includes investment in subsidized "public housing," which combines guaranteed private investment with public subsidies to bring this housing within reach of low-income families. The earlier chart on page 47 indicates that, from a 1962 base, employment in contract construction (of which employment in residential construction should be an increasingly large part) needs to be lifted 37.8 percent by 1966, and 48.6 percent by 1970.

The following chart illustrates this chapter.

# ROLE OF HOUSING IN THE U.S. ECONOMY 1947-1962

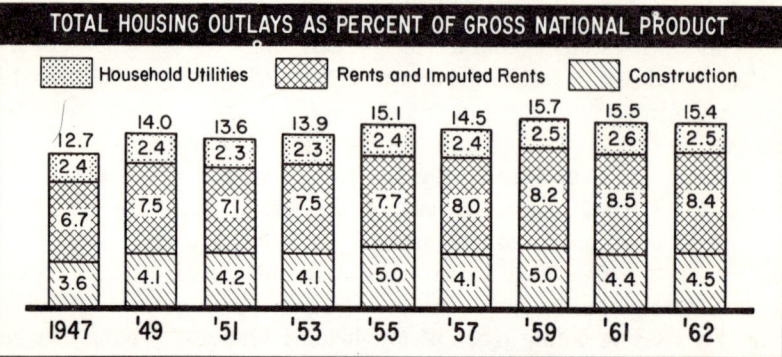

**TOTAL HOUSING OUTLAYS AS PERCENT OF GROSS NATIONAL PRODUCT**

Household Utilities — Rents and Imputed Rents — Construction

| | 1947 | '49 | '51 | '53 | '55 | '57 | '59 | '61 | '62 |
|---|---|---|---|---|---|---|---|---|---|
| Total | 12.7 | 14.0 | 13.6 | 13.9 | 15.1 | 14.5 | 15.7 | 15.5 | 15.4 |
| Household Utilities | 2.4 | 2.4 | 2.3 | 2.3 | 2.4 | 2.4 | 2.5 | 2.6 | 2.5 |
| Rents and Imputed Rents | 6.7 | 7.5 | 7.1 | 7.5 | 7.7 | 8.0 | 8.2 | 8.5 | 8.4 |
| Construction | 3.6 | 4.1 | 4.2 | 4.1 | 5.0 | 4.1 | 5.0 | 4.4 | 4.5 |

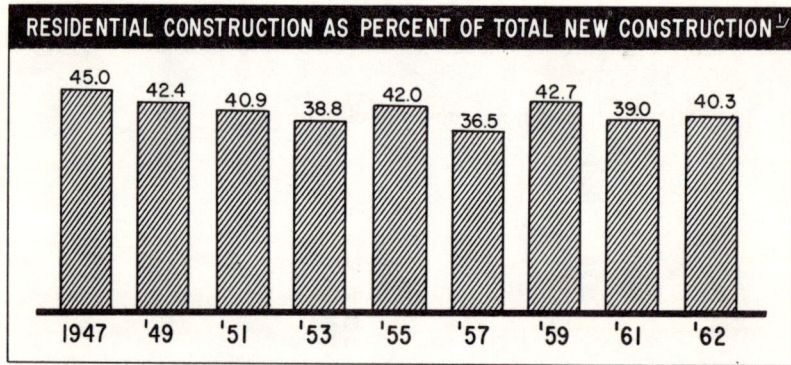

**RESIDENTIAL CONSTRUCTION AS PERCENT OF TOTAL NEW CONSTRUCTION** [1]

| 1947 | '49 | '51 | '53 | '55 | '57 | '59 | '61 | '62 |
|---|---|---|---|---|---|---|---|---|
| 45.0 | 42.4 | 40.9 | 38.8 | 42.0 | 36.5 | 42.7 | 39.0 | 40.3 |

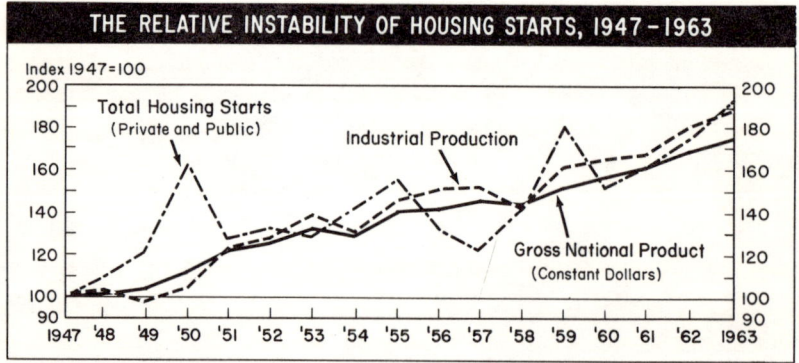

**THE RELATIVE INSTABILITY OF HOUSING STARTS, 1947-1963**

Index 1947=100

Total Housing Starts (Private and Public)

Industrial Production

Gross National Product (Constant Dollars)

[1] Private and public, nonfarm and farm, inc. additions and alterations.

Data: Dept. of Commerce, F.H.A., and V.A.

# XI. Social Need
# For Greatly Expanded Housing Efforts

### Extent of bad housing in the U. S., and the outlook

From the basic viewpoint of producing more of what the Nation really lacks, the claim of housing is extraordinarily high. The overwhelming majority of Americans are reasonably well fed and clothed, despite serious deficiencies among the one-fifth of the Nation living in poverty. But a very large percentage of the American people are miserably housed. Here, our absolute standards and rate of progress have fallen far below what has been accomplished in some of the Western democracies overseas.

There is no simple litmus paper test as to what constitutes unsatisfactory housing. Structure, facilities, the availability of air and light, ease of accessibility, overcrowding, and even to a degree the setting, should all be taken into account. It seems that Census data tend to understate the housing problem in the United States. Nonetheless, as the chart on page 61 shows, 9.3 million housing units as of 1960 were designated by the Census as "seriously deficient," coming to about 16 percent of the total of 58.3 million units (allowing for Census understatement, at least a fifth of all Americans are ill-housed). This study estimates that 5 million of these seriously deficient units required replacement, and 4.3 million needed rehabilitation.

Projecting likely building trends without drastic changes in national housing programs and policies, this study estimates that 7.3 million housing units, or well above 11 percent of the total of 64.4 million units, would remain seriously deficient in 1966; and 5.2 million units, or more than 7.5 percent of the total of 68.9 million, would remain seriously deficient even by 1970. We should not tolerate the prospect that more than one American family in every 14 (again allowing for the Census understatement, more likely one in every 11) will remain in seriously deficient housing even by 1970. In addition, the quantitative shortage (resulting in overcrowding) of about 1¼ million dwelling units in 1960 would rise to 2.3 million dwelling units by 1966, and be 2.1 million even in 1970.

### Goals for improved housing conditions

The lower section of the same chart depicts, by way of contrast, the progress which would be made if the goals set forth in this study were to be achieved. These goals are entirely compatible with and indeed are a vital part of the overall goals for restoration and maintenance of full

59

employment and production.

By 1966, the seriously deficient dwelling units (as measured by the Census) would be reduced to 5.8 million, or about 9 percent of the total of 65 million, with 3.5 million requiring replacement and 2.3 million requiring rehabilitation. By 1970, the seriously deficient dwelling units would be reduced to 2.5 million, or only slightly above 3.5 percent of the total of 70.8 million, with 1.5 million requiring replacement and 1.0 million requiring rehabilitation. The quantitative shortage would be reduced to 1.5 million by 1965, and would be negligible by 1970.

## The human and public costs of bad housing

The social evils of bad housing are so visible and abundant that they need not be dwelt upon at great length. But the chart on page 62 gives some indication of the intimate relationship between substandard housing and the incidence of disease, juvenile delinquency and adult crime, accidents, and fires. The same chart depicts the relative density of the population in the so-called "blighted" areas, their small per capita contribution to tax revenues, and their relatively high drain upon municipal expenditures.

The two following charts are illustrative.

# TOTAL NUMBER OF HOUSING UNITS
# AND NUMBER SERIOUSLY DEFICIENT
# IN 1960 AND PROJECTED TO 1966 & '70

### ( Millions of Dwelling Units )

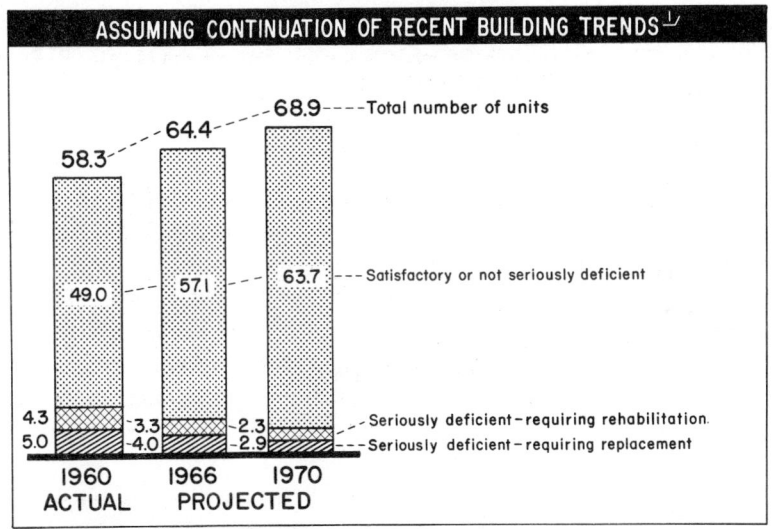

ASSUMING CONTINUATION OF RECENT BUILDING TRENDS[1]

68.9----Total number of units
64.4
58.3

49.0 --- 57.1 --- 63.7 ---Satisfactory or not seriously deficient

4.3
5.0
-3.3
-4.0
-2.3
-2.9
Seriously deficient – requiring rehabilitation.
Seriously deficient – requiring replacement

1960    1966    1970
ACTUAL   PROJECTED

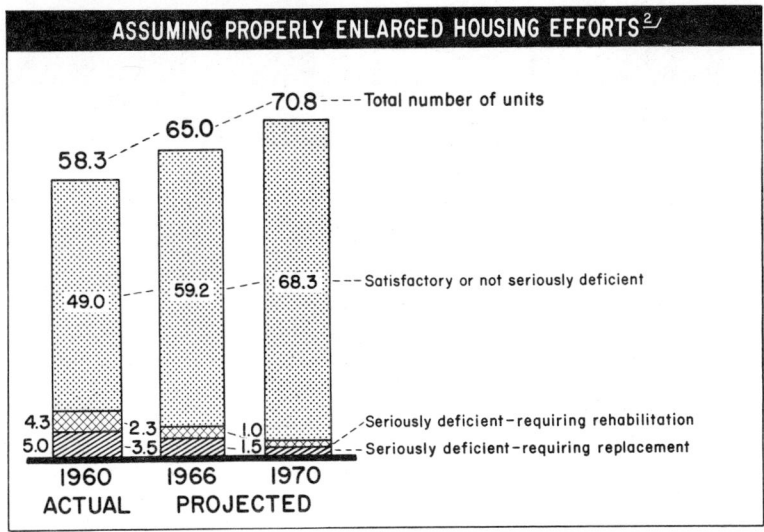

ASSUMING PROPERLY ENLARGED HOUSING EFFORTS[2]

70.8----Total number of units
65.0
58.3

49.0 --- 59.2 --- 68.3 ---Satisfactory or not seriously deficient

4.3
5.0
2.3
-3.5
1.0
-1.5
Seriously deficient – requiring rehabilitation
Seriously deficient – requiring replacement

1960    1966    1970
ACTUAL   PROJECTED

[1] In addition,the quantitative shortage was 1.25 million units in 1960, and would be 2.3 million in 1966 and 2.1 million in 1970.
[2] The quantitative shortage would be 1.5 million units in 1966 and negligble in 1970.

1960 Data: U. S. Census Bureau

61

# SUBSTANDARD HOUSING BREEDS
# ECONOMIC AND SOCIAL ILLS

## SACRAMENTO, CAL.

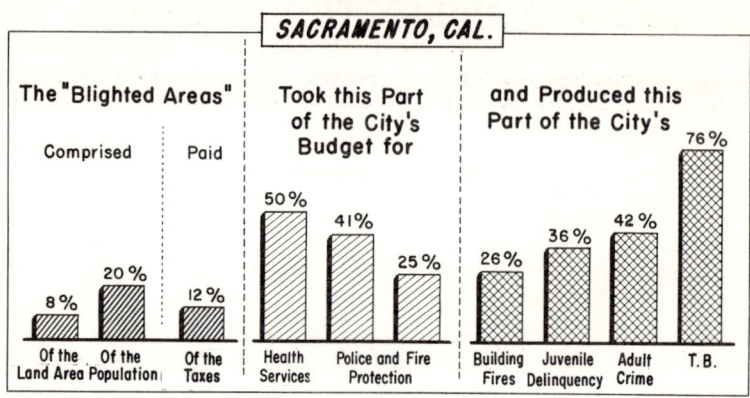

The "Blighted Areas"

Comprised | Paid

8% Of the Land Area
20% Of the Population
12% Of the Taxes

Took this Part of the City's Budget for

50% Health Services
41% Police and Fire Protection
25%

and Produced this Part of the City's

26% Building Fires
36% Juvenile Delinquency
42% Adult Crime
76% T.B.

Source: Sacramento Planning Commission

## LOS ANGELES, CAL.

Comparing "Blighted Areas" with a Control "Good Area"
(On a Per Capita Basis)

For every Tax Dollar | The "Blighted Areas" Paid | For every Tax Dollar | The "Blighted Areas" Cost

$1 — Paid by the "Good Area"

38¢ in Taxes

$1 — Spent in the "Good Area"

$1.87 for Police Services
$1.67 for Fire Dept. Services
$2.25 for Health Services

Source: California State Commission of Housing

## LOUISVILLE, KY.

Comparing a Substandard Area with a Control "Good Area"
(With the same Population)

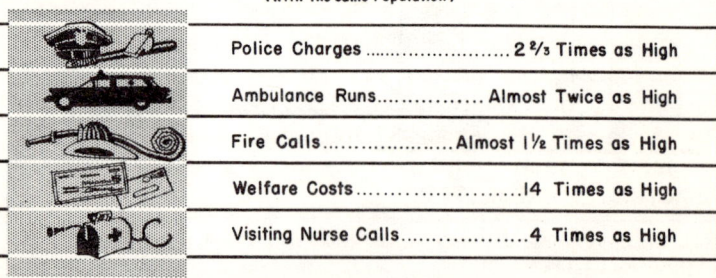

Police Charges .......................... 2 ⅔ Times as High

Ambulance Runs................ Almost Twice as High

Fire Calls..................... Almost 1½ Times as High

Welfare Costs........................14 Times as High

Visiting Nurse Calls....................4 Times as High

Source: Louisville Municipal Housing Commission

# XII. Bad Housing Results From Low Incomes

## Unsound housing related to income levels

The chart on page 65 depicts what portion of housing in metropolitan areas in 1960 was "unsound" according to Census classification. While all "unsound" housing is not as bad as the "substantially deficient" housing discussed in the previous chapter, the classification nonetheless serves to depict the relationship between bad housing and low income. With 20.9 percent of all rented dwelling units unsound, 28.9 percent of those rented by occupants with incomes under $4,000 and 18.9 percent of those rented by occupants with incomes between $4,000 and $6,000 were unsound. Only 7.3 percent of those rented by occupants with incomes over $10,000 was unsound. Stated in another way, 60.3 percent of the unsound housing was occupied by those with incomes below $4,000 and another 22.6 percent by those between $4,000 and $6,000, while only 2.9 percent was occupied by those with incomes of $10,000 and over. Owner-occupied housing is more largely for higher income people. But even here, with only 7.6 percent of the total number of dwelling units being unsound, 16.4 percent of those whose occupants were below $4,000 in income and 8.9 percent of those whose occupants were between $4,000 and $6,000 were unsound. Correspondingly, 47.5 percent of the owner-occupied unsound housing was lived in by those below $4,000, and another 23.4 percent was lived in by those between $4,000 and $6,000.

## Housing conditions related to housing costs

In metropolitan areas in 1960, as shown by the chart on page 66, 45.1 percent of all rental housing occupied by those paying gross monthly rents of less than $60 and 20.6 percent of that occupied by those paying between $60 and $80 were unsound, contrasted with only 6.6 percent of that occupied by those paying gross monthly rents of $100 and over. Correspondingly, 53.7 percent of the unsound housing was occupied by those paying less than $60, and another 27 percent by those paying between $60 and $80. In the case of owner-occupied housing, 27.4 percent of the dwelling units valued under $7,500 was unsound, and 58.6 of all the unsound units were under $7,500.

Occupants paying less than $60 a month or less than $720 a year in rent would be likely on the average to have annual incomes of less than $3,000, using a ratio of income to rent of about 4 to 1 (among those with such low incomes, the ratio is probably lower). More than 21 percent of all consumer units (multiple-person families and unattached individuals) in the United States have annual incomes below $3,000. Thus it appears from the data discussed above that, among the lowest fifth of the population, 4½ consumer units out of every 10 live in substandard housing. And among the one-eighth of all consumer units who have incomes under $2,000, probably at least 4 out of 5 live in unsound housing.

The two following charts provide more details.

# HOUSING CONDITIONS RELATED TO INCOMES IN METROPOLITAN AREAS, 1960

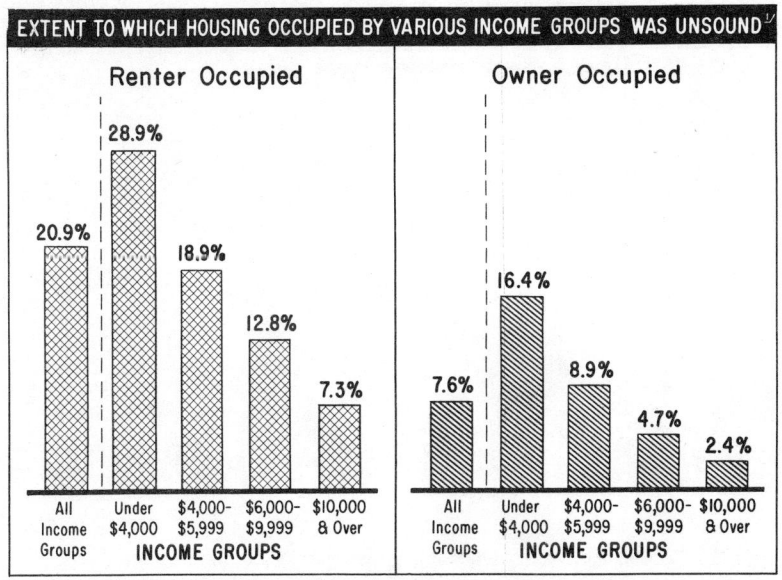

**EXTENT TO WHICH HOUSING OCCUPIED BY VARIOUS INCOME GROUPS WAS UNSOUND** [1]

### Renter Occupied

28.9%
20.9%
18.9%
12.8%
7.3%

| All Income Groups | Under $4,000 | $4,000-$5,999 | $6,000-$9,999 | $10,000 & Over |

INCOME GROUPS

### Owner Occupied

16.4%
7.6%
8.9%
4.7%
2.4%

| All Income Groups | Under $4,000 | $4,000-$5,999 | $6,000-$9,999 | $10,000 & Over |

INCOME GROUPS

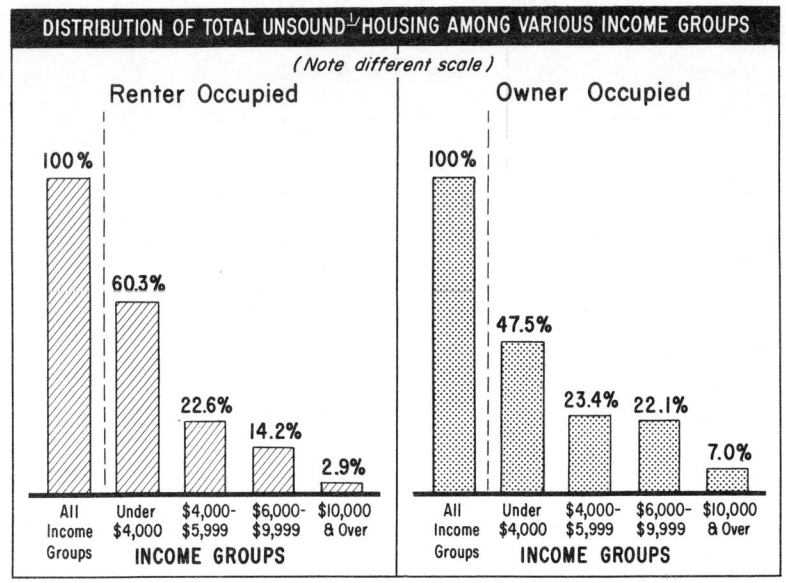

**DISTRIBUTION OF TOTAL UNSOUND** [1] **HOUSING AMONG VARIOUS INCOME GROUPS**

*(Note different scale)*

### Renter Occupied

100%
60.3%
22.6%
14.2%
2.9%

| All Income Groups | Under $4,000 | $4,000-$5,999 | $6,000-$9,999 | $10,000 & Over |

INCOME GROUPS

### Owner Occupied

100%
47.5%
23.4%
22.1%
7.0%

| All Income Groups | Under $4,000 | $4,000-$5,999 | $6,000-$9,999 | $10,000 & Over |

INCOME GROUPS

[1] Unsound housing is a Census classification based on some defects, but all unsound housing is not seriously deficient.

Data: U.S. Census of Housing, 1960

# HOUSING CONDITIONS RELATED TO COSTS IN METROPOLITAN AREAS, 1960

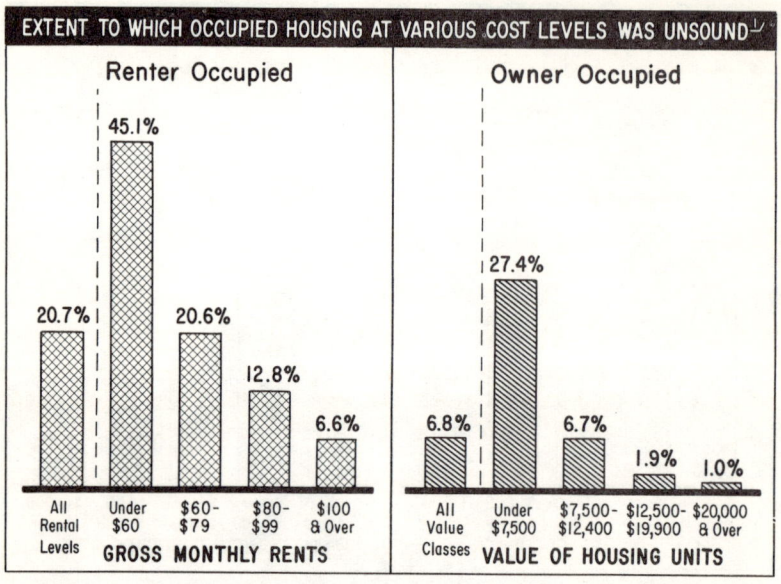

**EXTENT TO WHICH OCCUPIED HOUSING AT VARIOUS COST LEVELS WAS UNSOUND** [1]

### Renter Occupied

- All Rental Levels: 20.7%
- Under $60: 45.1%
- $60–$79: 20.6%
- $80–$99: 12.8%
- $100 & Over: 6.6%

GROSS MONTHLY RENTS

### Owner Occupied

- All Value Classes: 6.8%
- Under $7,500: 27.4%
- $7,500–$12,400: 6.7%
- $12,500–$19,900: 1.9%
- $20,000 & Over: 1.0%

VALUE OF HOUSING UNITS

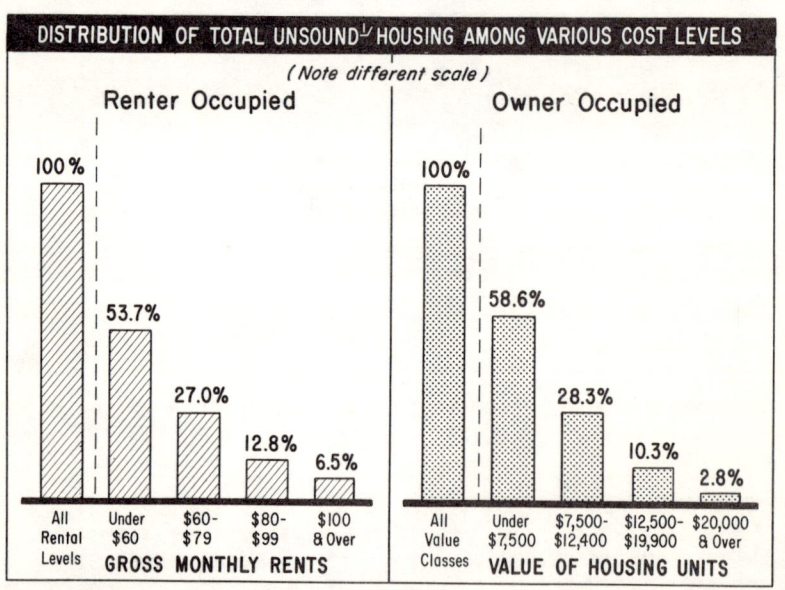

**DISTRIBUTION OF TOTAL UNSOUND** [1] **HOUSING AMONG VARIOUS COST LEVELS**

*(Note different scale)*

### Renter Occupied

- All Rental Levels: 100%
- Under $60: 53.7%
- $60–$79: 27.0%
- $80–$99: 12.8%
- $100 & Over: 6.5%

GROSS MONTHLY RENTS

### Owner Occupied

- All Value Classes: 100%
- Under $7,500: 58.6%
- $7,500–$12,400: 28.3%
- $12,500–$19,900: 10.3%
- $20,000 & Over: 2.8%

VALUE OF HOUSING UNITS

[1] Unsound housing is a Census classification based on some defects, but all unsound housing is not seriously deficient.

Data: U.S. Census of Housing, 1960

66

# XIII. Toward A Decent Home
# For Every American Family

## Why the need is not now being met

The task of lifting housing construction during the years immediately ahead to levels consistent with housing's role in the expansion of employment and investment opportunity set forth in Chapter X, and consistent with the goals for improved housing standards set forth in Chapter XI, cannot await the time when the incomes of the two-fifths of the Nation who now live in poverty or deprivation rise sufficiently for them to pay the costs of satisfactory housing under prevailing and forseeable housing costs. Instead, the expanded housing efforts which would enlarge employment and economic growth are essential to the reduction of poverty and deprivation, which has been slowed to a snail's pace by the poor U.S. economic performance during the past decade.

The inadequate levels of housing construction result inevitably from the fact that the great preponderance of new housing construction is for middle- and upper-income groups. Contrary to an old but now thoroughly discredited theory, the movement of these families into new housing does not leave an adequate supply of decent housing available for those lower down in the income structure; if it did, the slums in such vast magnitudes would not still be with us. Moreover, the construction of new housing mainly for these preferred groups recurrently saturates the so-called housing market. This explains the extraordinary instability of housing starts, which in turn has had so large an impact upon the poor performance of the whole U. S. economy.

To illustrate, during first quarter 1963, as shown by the chart on page 71, only 0.7 percent of the new FHA single-family homes were purchased by those with incomes under $4,000 (coming to 31 percent of all consumer units in the U. S. in 1962); less than 15 percent were purchased by all those with incomes under $6,000 (52 percent of all U. S. consumer units were in this income category); and more than 85 percent were built for those with incomes of $6,000 and over. In second quarter 1963, only 1.7 percent of the loans for new home purchases with down payments under the G.I. program were for veterans with incomes below $3,600. Actually, during the past decade, the incomes of the poor and deprived have not nearly kept up with rising housing costs.

## The proposed new effort

In 1963, out of an estimated 1.6 million new nonfarm housing starts, almost all were for middle- and high-income families, and only a very small fraction were for lower-middle and low-income families. The goal proposed for 1966 is 2 million nonfarm housing starts, with 1.2 million units for middle- and high-income families, 0.4 million for lower-middle-income families, and 0.4 million for low-income families. The goal for 1970 is 2.2 million nonfarm housing starts, with allocation among the three categories just mentioned of 1.2 million, 0.5 million, and 0.5 million. These goals are shown by the chart on page 72. In addition, the projected expansion of housing investment, earlier set forth, is far in excess of the projected expansion of housing starts, because it involves very large expansion of rehabilitation efforts.

The proposed traditionally financed private housing for middle- and high-income families does not require much change in available programs and policies. The proposed privately financed and cooperative housing for lower-middle income groups requires a vigorous admixture of private and public efforts: large increases in public outlays for aid to land acquisition and other aspects of urban renewal, which are essential to the housing effort itself; complete reversal of the tight money and high interest rate policy, which denies needed funds and adds to costs; and a variety of Federal loan or guaranty or insurance programs, going far beyond what the F.H.A. has thus far attempted in volume, with much lower interest rates and much longer terms of amortization in order to reduce annual rental costs and the annual costs of home ownership.

The needed expansion of the subsidized program for low-income families depends of course upon increasing efforts on the part of the States and municipalities. But it depends above all upon huge expansion in translating the Federally-assisted effort from a mere token program to the needed volume. This does not mean that the precise formula of aid which has been used to date should be adhered to in future, or that there is not opportunity to combine rental housing with a considerable amount of home ownership even for low-income people.

## The moderate Federal costs in terms of the benefits

The estimated Federal costs of this effort, plus the proposed costs of Federal assistance toward other aspects of urban renewal and community development, are not staggering, nor even large in the context of the Federal Budget and the benefits which would result. As shown by the chart on page 52 of a previous chapter, Federal Budget outlays for housing and community development should be lifted on a per capita basis related to the whole population from $1.42 in fiscal 1964 to $11.06 in calendar 1966 and $15.49 in calendar 1970. But measured in relationship to a properly growing American economy, these outlays would rise only from an indefensibly low .05 percent of our total national production in fiscal 1964 to the still very low ratios of 0.31 percent in calendar 1966 and 0.37 percent in calendar 1970. In absolute amounts, these Federal Budget outlays would rise from about a quarter of a billion dollars in fiscal 1964 to 2.2 billion in calendar 1966, and 3.3 billion in calendar 1970. These magnitudes do not seem large when compared with originally budgeted outlays in fiscal 1964 of more than 55 billion dollars for national defense, 4.2 billion for space research and technology, 5.7 billion for agriculture, and almost 2.7 billion for international affairs and finance.

## Relationship between housing and urban renewal

This discussion cannot deal in depth with aspects of urban renewal other than housing, although the foregoing goals for Federal outlays do include not only housing but also large expansion of Federal aid to other aspects of urban renewal. It should be stressed, however, that our nationwide urban renewal efforts—even while grossly inadequate—have run too far ahead of our housing efforts. In far too many cases, public subsidies are being used to help land owners, real estate interests, and powerful business interests, with actual detriment to the poor who live in the slums. With the slums in which they have been living "cleared," but with a gravely inadequate expansion of decent housing within their means, these people are crowded into other slums and made to pay higher rents to boot because of the shortages thus created. City "beautification" is all to the good, but it cannot conceal the slums nor lighten the burdens of the people who live in them. Beyond this, a housing effort of the size and nature indicated would immensely expand on a sounder basis every aspect of urban renewal. It would check the flight from our cities which is confronting them with insuperable financial difficulties, and dividing America into two worlds—the affluent out-of-towners and the poor at the city's core.

## The farm housing problem

The housing program set forth above does not include the needed improvements in farm housing. It must be recognized that the farm population, poorer than others in their personal incomes and in their receipt of public services, suffers also from an abundance of bad housing. But experience indicates that the improvement of housing conditions on the farm usually finds its more appropriate frame of reference in the treatment of the whole farm problem rather than the treatment of the housing problem *per se*. This study does include goals for improved farm income and output. The task of improving the incomes and living standards of our farm people, second to none in point of priority, has been covered comprehensively in several previous studies.*

## Unusually high "multiplier" effect of Federal aid to housing and community development

As earlier indicated, every additional dollar of Federal aid to housing and community development has a higher "multiplier" effect than an additional Federal dollar spent for almost any other purpose. This aid helps to stabilize private investment in residential construction and to improve the standards of such construction. It also enlarges private production and sale of the variety of goods and services which enter into the home. This aid stimulates many other types of private investment to service the improved urban areas.

In these connections, it should be noted that the proposals in this study to lift Federal outlays for housing and community development to a level 1.9 billion higher in calendar 1966 and 3 billion higher in calendar 1970 than in fiscal 1964, and to extend non-subsidy aids to lower-middle-income housing, are within the framework of projected increases in private residential nonfarm construction to levels 13.7 billion higher in 1966 and 22.7 billion higher in 1970 than in calendar 1963. These goals, in turn, are essentially related to projected levels of private investment in total new construction, running 24 billion higher in 1966 and 39.5 billion higher in 1970 than in 1963. Federal aid to the rehousing of those living in slums also has a very high "multiplier" effect upon the health and morale and ambition of people. And all of these types of expansion add greatly to Federal tax revenues at any given tax rates. All of these factors should be taken into account, as those who shape the Federal Budget rise to their full responsibilities in dealing with the problems of unemployment and low economic growth and strive properly to get the largest return for every dollar spent in terms of the real wealth of nations and in terms of the conditions of the Federal Budget itself.

The two following charts complete this study.

* See the Conference studies *Food and Freedom, Toward a New Farm Program,* and *Full Prosperity for Agriculture.*

# INCOMES OF PURCHASERS OF NEW FHA SINGLE FAMILY HOMES, 1ST QUARTER 1963

## PERCENT OF PURCHASERS IN VARIOUS INCOME GROUPS

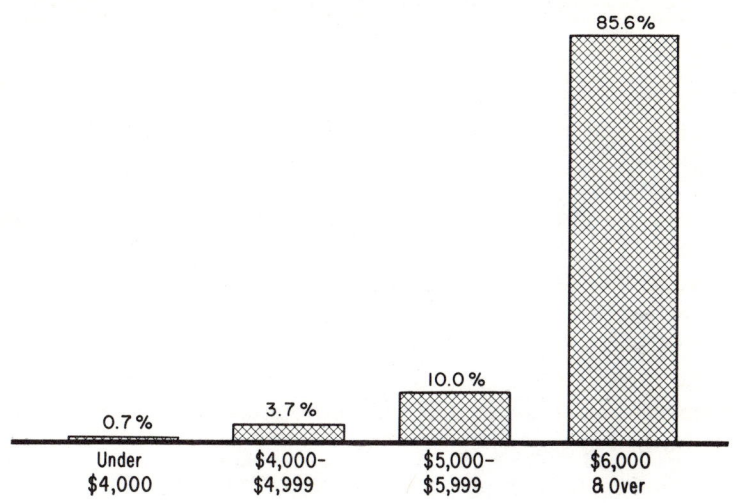

Data: Federal Housing Authority

# INCOMES (AFTER TAXES) OF HOME PURCHASERS UNDER G.I. HOME LOAN PROGRAM, 2ND QTR. 1963

## PERCENT OF VETERANS IN VARIOUS INCOME GROUPS

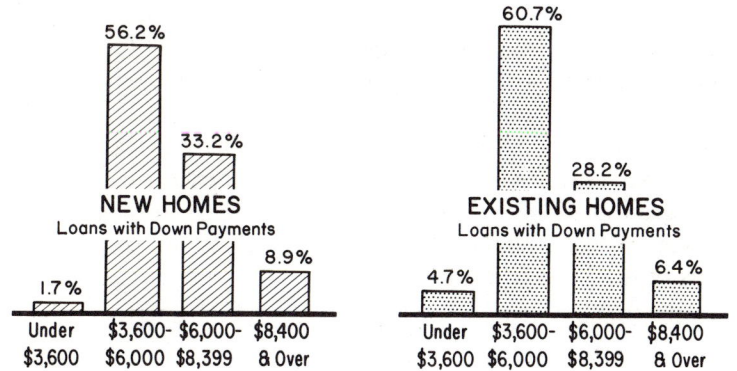

Data: Veteran's Administration

# TOWARD DECENT HOMES FOR ALL:
# GOALS FOR NEW NONFARM HOUSING

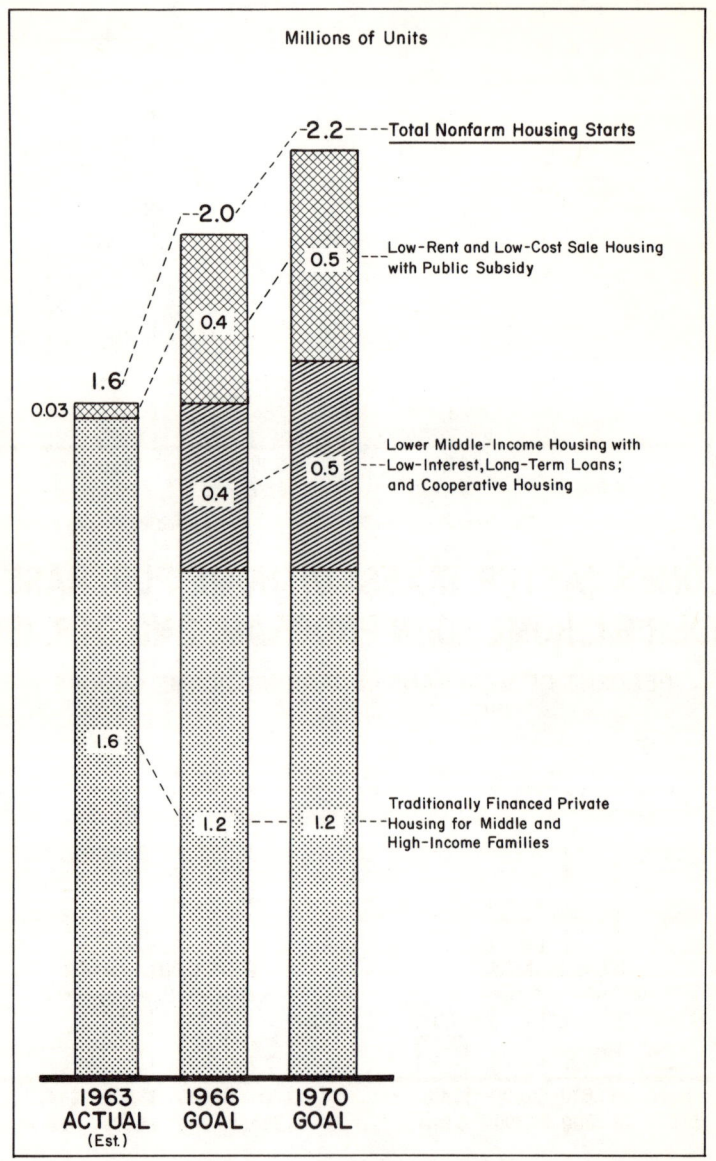

Millions of Units

2.2 ---- Total Nonfarm Housing Starts

2.0

0.5 ---- Low-Rent and Low-Cost Sale Housing
with Public Subsidy

0.4

1.6

0.03

0.4

0.5 ---- Lower Middle-Income Housing with
Low-Interest, Long-Term Loans;
and Cooperative Housing

1.6

1.2 --- 1.2 --- Traditionally Financed Private
Housing for Middle and
High-Income Families

1963
ACTUAL
(Est.)

1966
GOAL

1970
GOAL

1963 Data: Dept. of Commerce, F.H.A., and V.A.